Pore Pressure thı Mechanical Systems; the Force Balanced Physics of the Earth's Sedimentary Crust

by

Phil Holbrook Ph.D

© 2001 Force Balanced Petrophysics
All rights reserved.
Printed in the United States of America
ISBN: 0-9708083-0-5

For information, please contact:

Phil Holbrook, Force-Balanced.Net
2203 Blue Willow Drive.
Houston, TX 77042
Ph 713-977-7668
e-mail phil@Force-Balanced.net

ISBN 0-9708083-0-5

90000>

9 780970 808301

Pore Pressure through Earth Mechanical Systems

Preface

The earth's mechanical systems are composed of minerals and fluid. Porosity is the fluid–solid partitioning coefficient. Additionally its complement solidity (1.0-ϕ) is a definition of absolute *in situ* grain-matrix strain. The minerals and fluids of the earth have discrete density, elastic and grain-matrix-compactional stress/strain coefficients. These coefficients apply in several constitutive closed-form mineral-fluid mechanical system domains.

Earth mechanical systems are distinguished from all other pore pressure prediction methods in that they are;

1.) <u>Rigorously force balanced</u>;
2.) <u>Make direct constitutive correlations to *in situ* porosity and strain</u>; &
3.) <u>Are in direct accordance with Newtonian and Hooke's law physics.</u>

Each mechanical system involved in pore pressure calculation is closed-form and dimensionally correct. The Extended Elastic Equations relate V_p^2 and V_s^2 to mineralogy from 0 to 100% porosity. This mechanical system domain applies to static-elastic stress/strain, as well as dynamic-elastic borehole acoustic and seismic wave propagation.

The grain-matrix-compactional mechanical system applies from initial-grain-contact to total consolidation. The stress/strain and density coefficients of the five most common sedimentary minerals have been experimentally determined. NaCl brine compressibility and density coefficients are defined for all geologic PV/T conditions through an Equation of State. The five most common sedimentary minerals and NaCl brine compose over 90% of all sedimentary rocks.

The constitutive mineral & fluid, stress/strain & density, coefficient-mixing laws are linear within an earth mechanical system domain. Overburden, pore pressure and fracture propagation pressure are related to mineralogy and porosity in the grain-matrix-compactional mechanical system in ≈biaxial Normal Fault Regime basins.

The earth mechanical systems are straightforward in their application. Load, stress, and strain units are dimensionally correct and cancel within each mechanical system domain according to the governing physics.

The chapters of this book explain;

1.) The governing physics and definition of absolute in situ strain;
2.) The fundamentals of granular and clay mineral compaction;
3.) Compactional stress/strain relationships;
4.) Stress ratios evident in Normal Fault Regime and Strike-Slip basins;
5.) Mineralogy and fluid sensitive porosity calculation methods;
6.) The inter-related closed-form earth mechanical systems;
7.) Subsurface fracture pressure/pore pressure limits;
8.) High Pressure High Temperature fluid expansion calibration and prediction;
9.) Relationships to empirical pore pressure methods;
10.) The promise of earth mechanical systems.

There are bibliographies for each chapter and a cross-referenced glossary of technical terms. There are bookmarks and hyperlinks relating chapters, figures, tables, and equations to the glossary of technical terms. These digital shortcuts can quickly familiarize a reader that is not familiar with all the constitutive – mechanical relationships that are inherent in sedimentary rocks.

Earth mechanical systems are a new conceptual approach to subsurface geology and geologic engineering. Earth mechanical systems are broadly applicable to well-planning, drilling, casing, completion, geotechnical and reservoir engineering. Each of these subsurface engineering applications uses elastic, density, porosity, v:H:h loads, effective stresses, and pore pressure. These parameters are physically linked to each other through the earth constitutive mechanical systems.

The linkage of the grain-matrix compactional and Extended-Elastic-Equations mechanical systems domains are illustrated using diagrams and tables in Chapter 6.

Pore Pressure through Earth Mechanical Systems

Pore Pressure through Earth Mechanical Systems

Pore Pressure through Earth Mechanical Systems

Pore Pressure through Earth Mechanical Systems

Pore Pressure through Earth Mechanical Systems

Chapter   Table of Contents   page

Pore Pressure through Earth Mechanical Systems

Chapter   List of Tables   page

Pore Pressure through Earth Mechanical Systems

Pore Pressure through Earth Mechanical Systems

Pore Pressure through Earth Mechanical Systems

Chapter-1.0   THE GOVERNING PHYSICS OF GEOPRESSURE IN THE SUBSURFACE

The earth's interior is a closed physical-mechanical system. Loads in the subsurface are both generated and borne by minerals and fluid. Mineral and fluid density coefficients are combined to determine earth *in situ* vertical load. **Newton's Universal Law of Gravitation** applies to every earth element. The force of gravity is nearly constant within the uppermost 20,000 feet of the earth's sedimentary crust. In the sedimentary crust, **Newton's Law of Gravitation** can be expressed as a summation of bulk density or porosity partitioned mineral and fluid density coefficients;

$$S_v = \sum_0^{depth} \rho_b = [(\rho_M \bullet (1.0 - \phi) + (\rho_F \bullet \phi)] \quad \text{Equation-1.1}$$

The gravitational load (S_v) can be calculated as integrated bulk density (ρ_b) from any set of petrophysical sensors that provide porosity (ϕ), pore fluid density (ρ_F), and average mineralogy.

The mineral-grain matrix framework and pore fluid share the whole load. The Effective Stress Theorem, equation-1.2 is a fundamental force conservation law for fluid-filled granular solids (Carroll, M.M., 1980);

$$P_p = S_{ave} - \sigma_{ave} \quad \text{Equation-1.2}$$

In this physical - mathematical expression; pore fluid pressure (P_p) is the difference between the average external confining load (S_{ave}) and the average internal effective stress load (σ_{ave}).

Chapter-1.1 Solidity; solid-fluid partitioning coefficient **and** definition of compactional grain-matrix *in situ* strain

The key to *in situ* determination of load, stress, and fluid pressure relationships is that the rock property solidity (1.0 - ϕ) is also a definition of *in situ* grain-matrix compactional strain. Solidity (1.0 - ϕ) is equivalent to absolute *in situ* grain-matrix strain if no solid matter is added during burial compaction. Even if matter is added to or subtracted from a sedimentary rock, it's altered mineralogy is the effective stress load-bearing element. If the altered average mineralogy is known, loading limb effective stress can still be estimated from mineralogy and strain.

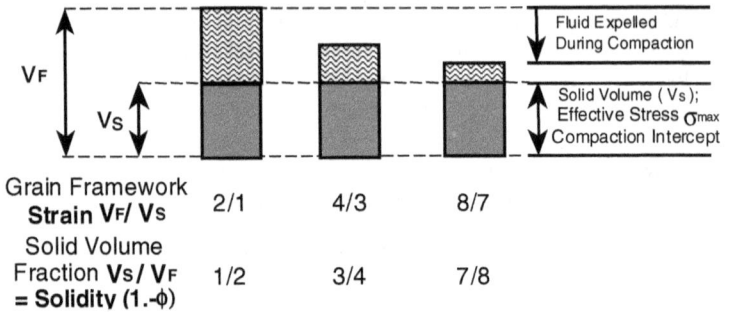

Solidity (1.− φ) is both the solid-fluid partitioning coefficient and a definition of volumetric grain-matrix framework strain for compacting granular solids.

Grain Framework Strain V_F/V_S	2/1	4/3	8/7
Solid Volume Fraction V_S/V_F = Solidity (1.-φ)	1/2	3/4	7/8

Figure-1.1 Relationship between solidity and grain-framework-strain for compacting granular solids. Strain is both axial and volumetric when the sides of the earth element have fixed dimensions. Solidity = 1.0 is the limit for grain matrix framework compaction. Adapted from Baldwin & Butler (1985).

Figure-1.1 illustrates the relationship between solidity and compactional strain assuming no change in solid volume during compaction. At solidity = 1.0, there is no porosity and the total external load is borne by the ionic bonds of the rocks mineral constituents. The effective stress load required to reduce a rock of a given mineralogy to zero porosity defines σ_{max} which is a mineral-grain-matrix physical property.

Force balance for the mineral-grain matrix framework in the earth is vectorial within volumetric. Volumetric effective stress is therein directly related to volumetric in situ strain i.e. (solidity). In this earth closed-form force↔balanced mechanical system, all the static solid vectors and the fluid scalars individually and collectively sum to zero. The closed-form solution involves only the density and grain-matrix-compactional coefficients of the minerals and fluid, that generate and bear the external loads.

Chapter-1.2 Compaction of mineralogic end-member sedimentary rocks

Sedimentary rocks are composed almost entirely of a few simple minerals. Mineralogy is the compositional control over sedimentary grain-matrix-compaction throughout its burial history. After properly accounting for force balance, only two mineralogic coefficients, (α & σ_{max}) are used to relate effective stress to grain-framework-strain. These are intrinsic sediment grain and rock properties.

A rock's final compaction resistance (σ_{max}) is primarily controlled by the rock's average mineralogic hardness. The effective stress compaction exponent (α) is related to mineral hardness and clay mineral inter-particle repulsive force. The relationships of these coefficients to the grain matrix will be discussed extensively in chapter 2. The first fundamental grain-matrix compactional *in situ* stress/strain relationship is;

$$\sigma_{ave} = \sigma_{max}(1.0 - \phi)^{\alpha} \quad \text{(Equation-1.3)}$$

This mineralogically general loading-limb stress/strain relationship was determined directly from *in situ* petrophysical measurements of five mineralogic end-members. The equation extends the shale compactional relationships of Baldwin & Butler (1985) to the other common sedimentary rocks. Holbrook(1995) demonstrated that the compaction curves of Gandino & Zennuchini (1987) for limestones, quartz sandstones, and shales are power-law *in situ* stress/strain functions. These power-law functions plus halite sand (Casas & Lowenstein, 1989), and anhydrite (Pfiefle & Senseny, 1981) are incorporated in figure-1.2.

Figure-1.2 shows the First Fundamental in situ Stress/Strain Relationship for natural single mineral granular sediments of the five most common sedimentary minerals (Holbrook, 1995). Four fixed composition minerals; quartz, calcite, anhydrite, halite and sedimentary clay minerals compose over 90% of all sedimentary rocks.

Sedimentary clay minerals are all composed of silica tetrahedral and slightly substituted alumina octahedral layers. These layers are similarly arranged with similar bond strengths. They all have negatively charged crystal faces and positively charged edges. The temperature and pressure factors that alter sedimentary minerals during burial diagenesis are incorporated into the average shale *in situ* measured compaction functions shown.

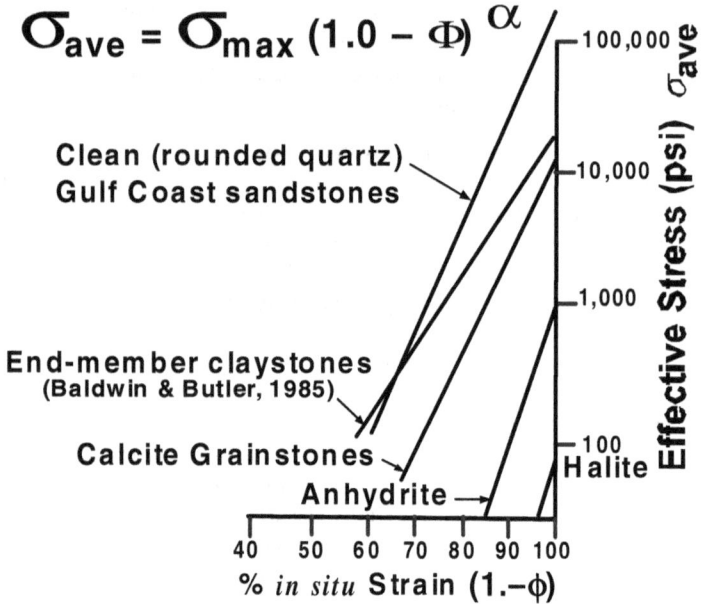

The First Fundamental *in situ* Stress/Strain Relationship for the five most common sedimentary minerals.

$$\sigma_{ave} = \sigma_{max} (1.0 - \Phi)^{\alpha}$$

Clean (rounded quartz) Gulf Coast sandstones

End-member claystones
(Baldwin & Butler, 1985)

Calcite Grainstones

Anhydrite

Halite

Effective Stress (psi) σ_{ave}

100,000
10,000
1,000
100

40   50   60   70   80   90 100

% *in situ* Strain (1.−ϕ)

Figure-1.2 The First Fundamental in situ Stress/Strain Relationship for the five most common sedimentary minerals. Average effective stress (σ_{ave}) is borne by mineral ionic bonds and inter-particle repulsion between clay mineral particles. Taken from (Holbrook, 1999)

The loading limb effective stress consolidation of the common sedimentary minerals, kaolinite, illite, and montmorillonite are almost identical when their different cation exchange capacities are accounted for (Nagaraj, & Murthy, 1983). Quartz, calcite, anhydrite and halite have negligible cation exchange capacities. These minerals resist compaction only with their internal crystalline bonds, which can range from very high to very low. When these minerals are mixed with sedimentary clay; the compaction resistance of the composite is the volume weighted average of its mineralogic constituents (Holbrook, P, 1995).

Table-1.1 lists the average loading limb effective stress compaction coefficients for those 5 common sedimentary rock forming minerals as well as their respective hardness and solubility (Holbrook, 1996). Considering both table-1.1 and figure-1.2, the hardest least soluble common sedimentary mineral, quartz, also has the highest σ_{max} compaction resistance 130000 psi. Halite, or rock salt is the softest most soluble common sedimentary mineral, and has the lowest σ_{max} compaction resistance of about 85 psi. Hardness, solubility, and compaction resistance (σ_{max}) are all power-law relationships, which are controlled by the average ionic bond strength of the respective minerals. Minerals with stronger bonds have higher hardness, and are more compaction resistant. Minerals with stronger ionic bonds are also less soluble. Ions that are held more tightly in the mineral lattice are less likely to be taken out of their lattice sites and into solution. The measured hardness, solubility and grain density of these minerals are available in many standard mineral reference books (Carmichael, 1982).

Table-1.1   Power-law compaction coefficients, hardness, and solubility for naturally sedimented single-mineral grainstones and end-member claystones.

Mineral (or rock)	Sigma sub max (psi) comp limit	alpha comp exp	hardness (mhos)	solubility (ppm)
Quartz sand	130000	13.849	7.0	6
End-Member Claystone	18461	9.348	3.+	20
Calcite sand	12000	13.637	3.	120
Anhydrite	1585	20.646	2.5	3000
Halite	85	32.564	2.0	350000

The power-law effective stress compaction coefficients (σ_{max} & α) on table 1.1 were derived from the following sources; Quartz sand (Atwater & Miller, 1965), Average Shale (Holbrook, P W, 1995a), Calcite Sand (Gandino & Zenucchini, 1987), Anhydrite (Pfiefle & Senseny, 1981) and halite Sand (Casas & Lowenstein, 1989). Force balance and constant hydrostatic pore pressure were applied to derive the coefficients for each dataset. Mineral hardness, solubility and density can be looked up in (Carmichael, 1982) or any Handbook of Chemistry and Physics.

Power-law relationships can naturally average multi-dimensional physical relationships. Loads are borne across grain contacts and through grains. The average load is borne across "n" grain contacts in all directions. A partial load is borne at each of "n" contacts in all directions. A power-law exponent (α) simultaneously captures the "n" exponential coordination number and particle interaction relationships that affect loading limb compaction.

Average load bearing is overall volumetric through mineral grains, non-clay grain contacts, and clay mineral inter-particle repulsive force. These individual power-law relationship will be discussed in chapter 2. These micro power-law relationships combine to yield the average mega power-law relationships shown on figure-1.2 and table-1.1.

References cited

Atwater, G. L. & E. E. Miller, 1965, "The Effect of Decrease in Porosity With Depth on Future Development of Oil and Gas Reserves in South Louisiana"[abs.]: AAPG Bulletin v. 49, p.334

Baldwin, B. & C.O. Butler, 1985, "Compaction Curves", AAPG Bulletin, Vol. 69, No. 4, pp. 622-626.

Carmichael, R.S., 1982, "Handbook of Physical Properties of Rocks", CRC Press.

Carroll, M.M., 1980, "Compaction of Dry or Fluid-filled Porous Materials", Journal of Engineering Mechanics Division, Proceedings of the American Society of Civil Engineers, Vol. 106, No EM5, Oct 1980 pp969 - 990.

Casas, E., & T. K. Lowenstein, 1989, "Diagenesis of Saline Pan Halite: Comparison of petrographic features of modern, Quaternary, and Permean Halites", Journal of Sedimentary Petrology, (USA) 59(5), pp. 724-739.

Gandino, A., & G. Zenucchini, 1987, "Density Depth Correlation in Po Valley Sediments." Bollettino de Geofisica Teorica ed Applicata, Vol XXIX, pp. 221-231.

Holbrook, P.W., 1995, "The relationship between Porosity, Mineralogy and Effective Stress in Granular Sedimentary Rocks", paper AA in **SPWLA 36th Annual Logging Symposium**, June 26-29, 1995.

Holbrook, P W, 1999, "A simple closed-form force balanced solution for Pore pressure, Overburden and the principal Effective stresses in the Earth.", **Journal of Marine and Petroleum Geology**, Vol. 16, pp. 303-319.

Newton, Isaac, 1687, Philosphiae Naturalis Principia Mathematica.

Terzaghi, K. Van, 1923, "Die Berchnung der Durchassigkeitziffer des Tones aus dem Verlauf der Hydrodynamischen Spannungscheinungen", Sitzunzsber Akad Wiss. Wein Math Naturwiss, K1.ABTS 2a, pp. 107-122.

Chapter-2.0 THE PRIMARY CONTROLS OVER SEDIMENT COMPACTION

Mineralogic composition is the primary control over sediment compaction. Near surface sediments and sedimentary rocks compact in proportion to the effective stress load applied to the grain-matrix framework. The load borne by the grain framework is a volumetric force balance relationship within and between solid particles. Mineral ionic bonds bear the load within particles and across direct grain-grain contacts. Electrostatic repulsive forces between negatively charged clay mineral particles also bears a portion of the effective stress load.

There is a strong correlation between average particle size and initial porosity of natural marine sediments. Compaction resistance is also strongly correlated to average particle size. Clay mineral inter-particle repulsive force explains both correlations. The particle size vs. compaction-resistance relationship is continuous from coarse sands to the finest particle size clays. Graded bedding due to individual particle settling velocity sorting places mineral grains of similar size and mechanical properties next to each other.

Five mineral specific compaction functions were determined from *in situ* petrophysical data. Average mineral ionic bond strength controls mineral hardness, solubility, and the plastic compaction intercept. Clay minerals have an additional inter-particle electrostatic repulsive force that is inversely proportional to sedimentary clay particle size and directly proportional to clay surface area in a given rock volume. All these factors are simultaneously accounted for by two power-law compaction coefficients (σ_{max} & α).

Solidity (1.0–porosity) is an end-of-plastic-compaction *in situ* strain parameter. The (solidity=1.0) intercept (σ_{max}) incorporates both elastic and plastic grain matrix strain into volumetric *in situ* strain. The power-law compaction exponent (α) captures the inter- and intra- particle compaction resistances mentioned above with respect to (σ_{max}).

Chapter-2.1 Structure, ionic bond strength, and charge distribution of the common sedimentary minerals.

Figure-2.1 shows the crystal lattice structures of the four most common sedimentary mineral types. The common non-clay minerals have a directionally neutral internal charge distribution. Positive and negative

8

charges are equal on the Angstrom scale within the mineral. Contacts between non-clay mineral particles are on the much larger micron scale and therefore have no net electrostatic charge. Adjacent non-clay mineral particles and those in direct contact will have no net electrostatic repulsive forces between them.

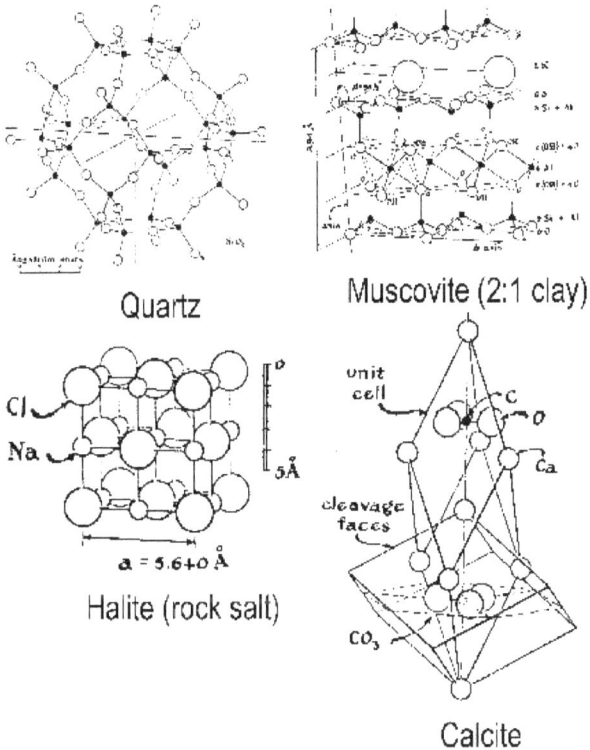

Figure-2.1. Crystal lattices of the four most common sedimentary mineral types. Quartz, muscovite (representative 2:1 clay), calcite, and halite (rock salt) are shown. All minerals have a net neutral internal charge balance. The non-clay minerals cleave in all directions presenting net neutrally charged mineral grain surfaces. The clay minerals cleave preferentially perpendicular to the c-axis presenting a negatively charged oxygen anion layer at the clay particle surface. Adapted from Berry & Mason (1959).

Sedimentary clay minerals all possess a layered structure of dominantly silicon centered tetrahedra and aluminum centered octahedra as shown on figure-2.1. The positively charged (+) aluminum, silicon and other high positive valence ions are sandwiched between negatively charged (-) oxygen layers. Clays cleave preferentially normal to their c-axis exposing large surfaces of negatively charged (-) oxygen anions. Sedimentary clay particles are thin platelets with negatively charged faces and positively charged edges. The negative/positive charge ratio over the entire surface of sedimentary clay mineral platelets is very large in the range of 2500:1 to 250000:1.

Chapter-2.2 Ionic bond force balance within mineral particles and across non-clay mineral grain contacts

Non-clay minerals resist compaction through their mineral lattice and at direct grain contacts with other non-clay minerals. A given particle is in contact with its neighbors over a fractional area of that particle. Particle contact area limits are from zero in a particle fluid suspension to 1.0 when intergranular porosity is reduced to zero.

Force balance across the fractional contact area is the product of the number and strength of mineral ionic bonds. Harder minerals with stronger ionic bonds such as quartz can bear a given load over a smaller area than soft minerals like halite. Particle contact area varies in proportion to volumetric solidity. Both particle area and solid volume reach 1.0 at the sedimentary rock's grain-matrix-compactional limit (σ_{max}).

Chapter-2.3 Direct non-clay mineral grain contact pressure solution compaction mechanism

Pressure solution is a major compaction mechanism on a geologic time scale. On a geologic time scale the mineral grain matrix tends to behave like an ideal plastic. Higher stress across neutrally charged grain-grain contacts is gradually equalized through pressure solution and generally local re-precipitation.

Mineral hardness and solubility in water are power-law relationships (Carmichael, 1982); so is single mineral particle compaction resistance (Holbrook, 1995a). All these chemical-mechanical attributes are fundamentally controlled by average mineral ionic bond strength.

Stress is everywhere proportional to strain in the earth's mechanical systems. The stress field is three-dimensional and the coordination number of adjacent solid load bearing grains is "n" dimensional. The loading limb and un-loading limb stress/strain relationships for porous granular solids are multi- dimensional i.e. power-law relationships.

Chapter-2.4 Compaction of non-clay granular sediments

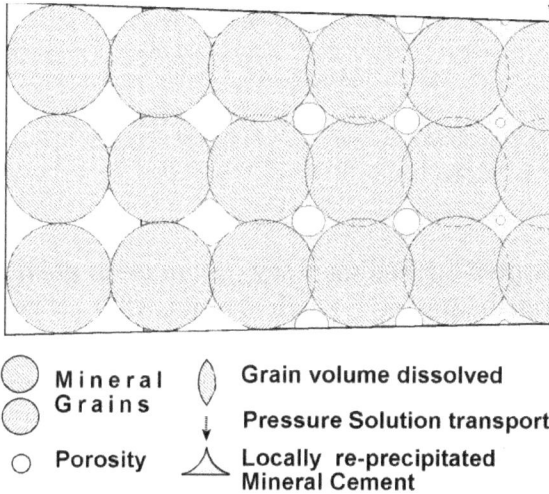

Mineral Grains	Grain volume dissolved
Porosity	Pressure Solution transport
	Locally re-precipitated Mineral Cement

Figure-2.2a shows a representative transition of neutrally charged spherical grains from initial grain contact to complete consolidation. The initial depositional porosity of rounded sedimentary grains is about 40 percent. The left border of the diagram represents the point of initial gravitational grain-grain contact. There is zero effective stress at initial contact. Physically this represents the surface of the earth overlain by fluid or air. The rightmost edge of figure-2.2 a is the plastic compaction limit (solidity = 1.0 @ σ_{max}).

Figure-2.2a represents solid volume conserved compaction of rounded neutrally charged mineral grains. As the grains are forced together, minerals are dissolved preferentially at the more highly stressed grain-grain contacts and re-precipitated in the nearby pore space. Net pore volume is indicated in white on figure-2.2a. Idealized pore shape would change progressively from an inverse sphere, toward small isolated spheres, when porosity reduction is accompanied by pressure solution followed by local re-precipitation.

Initial grain size is not reduced from left to right on figure-2.2a. The compaction portrayed is accomplished entirely by transporting mineral matter from higher stress grain-grain contacts into the available pore space. Natural quartz grains have different initial shapes. Mineral grains may be fractured, re-healed, and re-arranged during compaction.

Quartz cement has the same mineral lattice structure and the same load bearing capacity as the quartz grain from which it was dissolved. Through pressure solution the neutrally charged grain matrix increases the number of mineral ionic bonds at each grain-grain contact to support the average effective stress load. The remainder of the total load is borne by pore fluid pressure. While an agglomerated grain matrix has some finite porosity, the total confining load is shared between 1.) solid net neutrally charged mineral ionic bonds, 2.), inter-particle electrostatic repulsive forces, and 3.) pore fluid pressure.

Chapter-2.5 Sedimentary clay mineral particle size vs. electrostatic inter-particle repulsive force relationship

Most natural sediments and sedimentary rocks are a mixture of clay and non-clay minerals. Electrostatic inter-particle repulsion plays an important role in the compaction of these natural sediments. Figure-2.2b shows a microscopic representation of a low porosity clay dominated soil. The clay mineral packets shown (kaolinite, illite, and smectite) have similar particle size/shape aspect ratios. Clays are usually sedimented with varying percentages of neutrally charged silt grains also shown. The net compaction resistance of a naturally deposited sedimentary mixture is the mineralogically weighted average of the individual grain compaction resistances (Holbrook, 1995b).

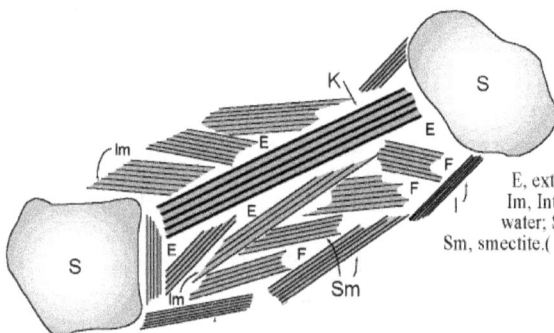

Figure 2b. Sedimentary particles and forms of water in a low porosity clay soil. F, free or bulk water; E, external or intercluster water; Im, Interlammelar or inter-cluster water; S, silt; K, kaolinite; I, illite; Sm, smectite.(taken from Hueckel, 1992)

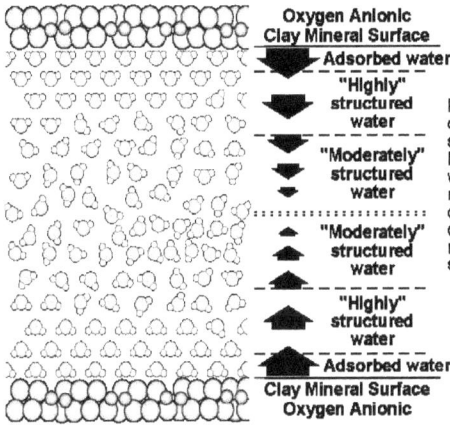

Oxygen Anionic
Clay Mineral Surface
Adsorbed water
"Highly"
structured
water
"Moderately"
structured
water
"Moderately"
structured
water
"Highly"
structured
water
Adsorbed water
Clay Mineral Surface
Oxygen Anionic

Figure 2c. Two opposing negatively charged clay mineral oxygen anionic surfaces shown at the atomic scale. Negative charge density deminishes with distance from the oxygen anionic mineral surface. Negative charge density both orients the nearby water dipole molecules and repels other negatively charged clay mineral surfaces

Figure-2.2c is a molecular scale representation of the space between two adjacent clay mineral particles. The water dipoles are held tightly to the electrostatically charged clay mineral oxygen anionic surface as shown. Water dipoles occilate more randomly as the negative charge density originating from the clay mineral surface falls to zero.

The different clay minerals shown on Figure-2.2 b, and Figure-2.2 c have important internal compositional differences. However, these different minerals have very similar micro-mechanical properties that are all related to particle size as indicated on figure-2.3. There is an essentially 1:1 power-law relationship between clay mineral Specific surface area and Cation Exchange Capacity. Clay mineral particle size decreases as surface area and CEC increase on Figure-2.3. All these laboratory clay minerals occur along the 1:1 line that is also a simple power-law particle-size relationship.

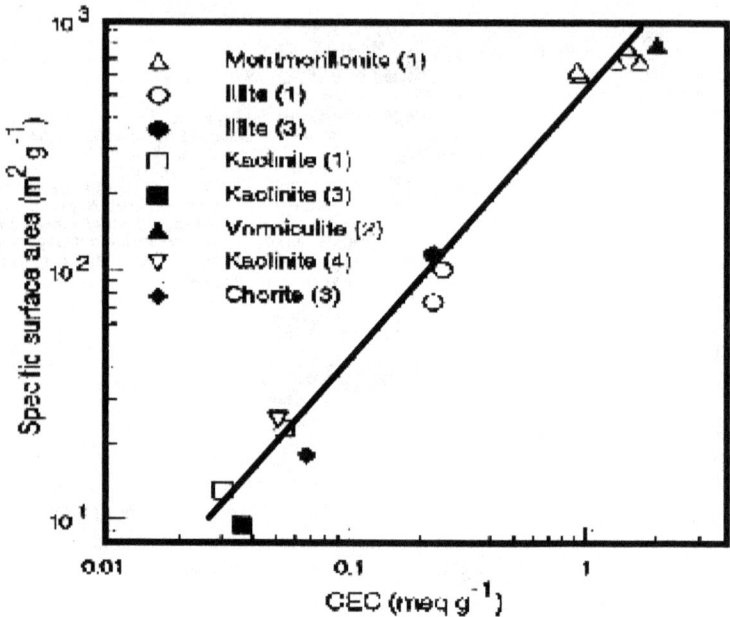

Figure-2.3, Specific surface area vs. Cation Exchange Capacity for various clay minerals. Taken from (Revil, et al, 1997). The strong correlation between surface charge area and Cation Exchange Capacity of sedimentary clay minerals is approximately a 1:1 areal constant. Repulsive forces between clay particles are directly proportional to negatively charged surface area and inversely proportional to clay mineral particle size. Sedimentary clay minerals are mechanically similar due to similarly high Al/Si ratios.

Both clay and non-clay mineral particle size at initial sedimentation is determined by particle settling velocity. The clay mineral CEC-particle size continuum shown on Figure-2.3 is an extension of the general Reynolds number particle size settling velocity continuum for all particles that are settled gravitationally.

The dominant underlying control over clay mineral electrostatic repulsive forces and mechanical properties is also clay mineral particle size. This relationship is set gravitationally at the moment of initial particle settling agglomeration at the earth's surface. Natural marine sediments with longer water columns best demonstrate the particle size, initial porosity, inter-particle repulsive force relationships.

Chapter-2.6 Mineral dependant Grain Size ↔ Inter-particle Repulsive Force ↔ Initial Porosity Relationships

The zero effective stress compaction starting point for granular sediments can be determined from porosity-mineralogy-particle size relationships in marine sediments at the sea floor (Skempton, 1970, Shumway, 1960). The approximately 10 cm below the surface line provides a sample that is close to zero effective stress. Figure-2.4 shows the inter-relationships between observed initial depositional porosity, laboratory measured liquid limit, and observed *in situ* compactional stress/strain relationships. Effective vertical stress is on the horizontal axis; and the two more common measures of strain, void ratio and porosity are shown on the vertical axis. Liquid limit is strongly correlated to CEC (Nagaraj & Murthy, 1983). Therefore the effective stress compactional relationships shown on Skempton's figure-2.4 are also particle size and inter-particle repulsive force dependent.

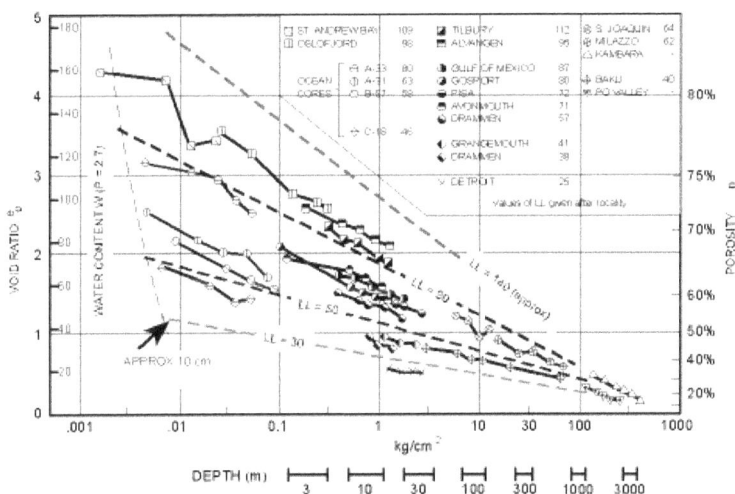

Figure-2.4 shows the inter-relationships between observed initial depositional porosity, laboratory measured liquid limit, and observed *in situ* compactional stress/strain relationships.

The grain density of quartz and clay are approximately equal and the effective stress gravitational load in the uppermost meter of the seafloor is very low. The difference between 80% porosity for fine clay deposits and 40% porosity for medium quartz sand deposits is due to the repulsive force between individual clay particles. The inter-particle repulsive force is weak but persistent with increasing effective stress compaction as indicated on Skempton's figure-2.4 . The figure-2.5. particle size crossplot coincides with the 10 cm line on figure-2.4.

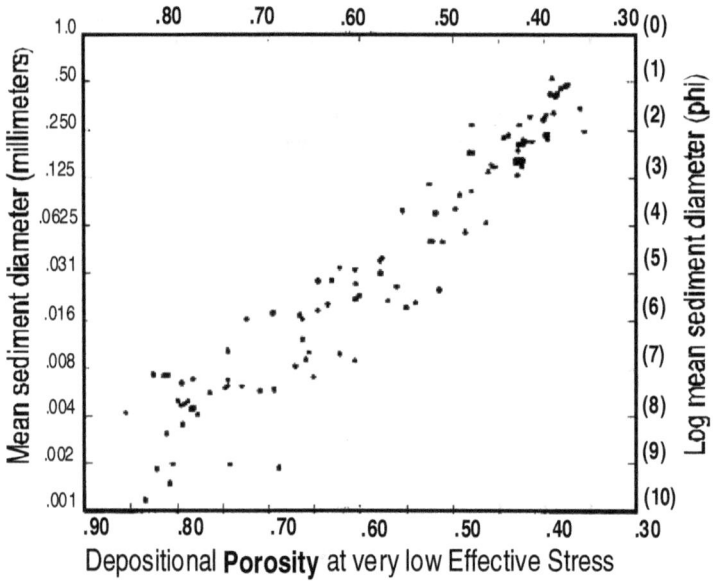

Figure-2.5. Mean sediment diameter vs. porosity for surface marine sediments taken from Shumway (1960). Depositional porosity varies uniformly from 40% for (0.5 millimeter) quartz sands to 90% for a very fine (0.001 mm) sedimentary clay. Porosity at very low effective stress is a function of particle size and clay mineral inter-particle repulsion.

Nagaraj & Murthy, (1983) later explained Skempton's entire compactional relationship mechanically in terms of negatively charged mineral surface area. Cation (+) Exchange Capacity (CEC) is essentially a direct measure of clay mineral negatively charged (-) oxygen anion surface area. The average electrostatic repulsive force between clay particles is also proportional to Cation Exchange Capacity. The compactional relationships of all clay minerals were normalized to the same curve based upon their average Cation Exchange Capacity.

Adjacent particles in a sedimentary layer have fallen through the same water medium at about the same speed. Clay particles in the sand size range will have proportionally lower external surface area and therefore lower average inter-particle repulsive force. Graded bedding places grains of like size and inter-particle repulsive force adjacent to each other. Particle size and negative surface charge area work together in the marine environment to generate the strong inter-dependent relationships shown on figure-2.5, figure-2.4, and figure-2.3.

Chapter-2.7 General Mineralogic Stress/Strain Loading for Granular Solids

Figure-1.2 showed the volumetric effective stress loading-limb compactional relationships for natural single mineral sedimentary deposits over the entire depth range of drilling interest. The loading-limb stress/strain relationships are global in nature dependent principally upon mineralogic composition (Holbrook, 1995a).

The effective stress compactional relationships were measured *in situ* taking overburden and pore pressure force balance into account (Terzaghi, 1923). The volumetric Effective Stress Theorem (Carroll, 1982) was also honored (P. Holbrook, 1999). These macro-mechanical stress/strain relationships are a composite of the micro-mechanical relationships (figure-2.3, figure-2.4 and figure-2.5) previously discussed. The composite of several mineral specific micro-scale power-law relationships results in a macro-power-law grain-matrix-compactional relationship (Figure-1.2).

The four neutrally charged non-clay minerals; quartz, calcite, anhydrite, and halite; have sub-parallel power-law loading limb effective stress/strain coefficients (α). These neutrally charged single mineral stress/strain relationships are offset from each other in proportion to their plastic compaction intercepts (σ_{max}). The compaction intercept (σ_{max}) is positively related to mineral hardness and inversely related to mineral solubility (table 1.1). All three are measures of the average inter-ionic bond strength of these sedimentary minerals.

Naturally sedimented clay minerals also have power-law loading-limb effective stress/strain coefficients (α) and a plastic compaction intercepts (σ_{max}). As shown on figure-1.2, the (σ_{max}) compaction intercept is well below granular quartz, and above granular calcite. This corresponds to clay mineral ordinal rankings between quartz and calcite on the hardness and solubility physical property scales. This supports the conclusion that (σ_{max}) is a mineralogic compactional stress/strain physical property.

The average sedimentary claystone effective stress/strain coefficients (α) is distinctly different from the four neutrally charged minerals, quartz, calcite, anhydrite , and halite. So long as clay minerals have water-wet surfaces, the additional electrostatic repulsive forces between adjacent clay particles are effective. This additional electrostatic repulsive force contributes to the higher stress/strain coefficients (α) observed in naturally sedimented clay sediments and claystones.

This clay mineral repulsive force – particle size effect is shown by (Skempton, (1970), figure-2.4, Shumway, (1960), figure-2.5, and rationalized by Nagaraj & Murthy, (1983). This relative mineralogic compactional relationship is also observed in corresponding non-force balanced depth vs. porosity relationships as shown on the facing page.

Chapter-2.8 The granular quartz vs. claystone gravitational compaction crossover

The porosities of gravitationally compacted granular quartz sediments and clay rich sediments cross over each other in virtually all subsiding basins. Figure-2.6 shows average quartz sandstone and shale porosity vs. depth relationships from a well in the Gulf Coast. Above 2000 feet on figure-2.6, clay rich sediments have higher porosity than quartz sands. Below 2000 feet each of the two mineralogic end-members continue along their own compaction gradients and the curves diverge with increasing depth. Each mineralogic end-member follows its own smooth continuous compaction vs. depth trend throughout the burial history shown.

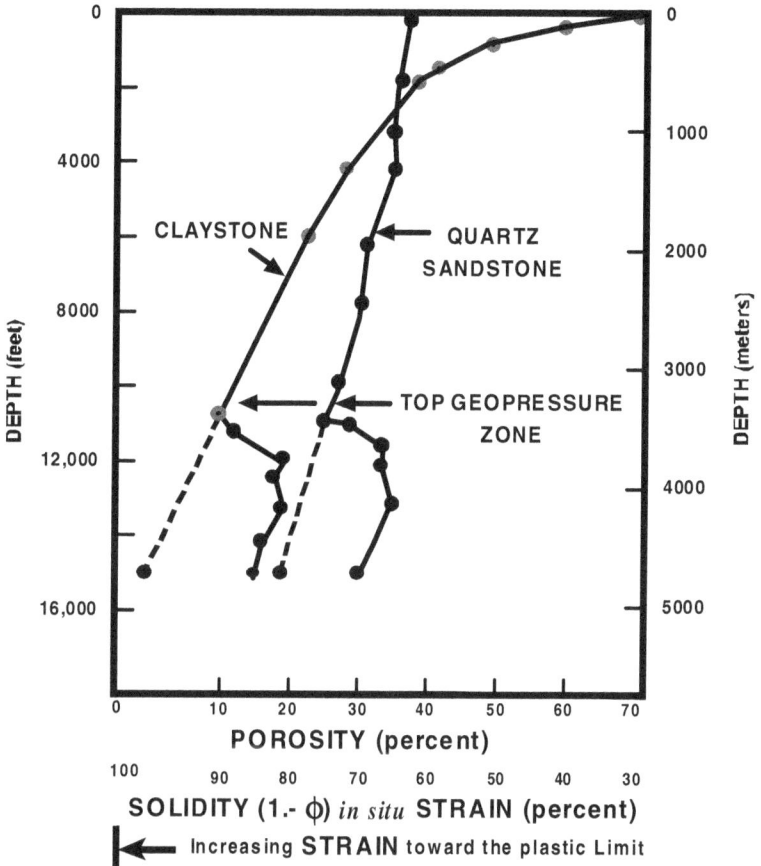

Figure-2.6 Quartz sand and shale porosity vs. depth compaction functions in the Gulf Coast. Taken from Stuart (1970). End-member-claystones have higher (α) plastic compressibility than quartz sandstones. The two mineralogic end-member *in situ* compaction curves cross at about 1000 feet.

Figure-1.2 has a corresponding mineralogic effective stress/strain crossover at about 300 PSI and 35% porosity. With increasing effective stress the quartz grainstone and worldwide claystone compactional stress/strain relationships diverge as they do on figure- 2.6. Skempton's figure-2.4 data also indicates a convergence in this same effective stress/strain region. The effective stress data on figure-2.4 and figure-1.2 are force balanced stress/strain relationships. Compactional porosity vs. depth data such as figure-2.6, correspond to the convergence and crossover shown on the effective stress functions of figure-2.4 , and figure-1.2.

This crossover occurs because of the claystone inter-particle repulsive force. Quartz grainstones settle to the seafloor with about 40% porosity or less. Depending on particle size and related inter-particle repulsive force, clay particles settle with an initial porosity of 75 to 95% porosity. The individual quartz grains are much harder than clay minerals and have high compaction resistance. Claystone electrostatic inter-particle repulsion and the minerals themselves are softer by comparison and therefore compact more easily. The primary controls over sediment compaction are these mineral specific physical properties.

If effective stress were properly accounted for on figure-2.6, the claystone points would plot on the worldwide claystone stress/strain power-law relationship whether they are "overpressured" or not. The same would be true of the quartz grainstone data points. The dashed line extensions of compaction trends on Figure-2.6 are entirely speculative. They take no account for the changing overburden load conditions nor do they account for the load borne by pore pressure. There is no mechanical sense in plotting depth, a measure of length against porosity.

The power-law effective stress and strain axes of figure-2.6 account for sediment compaction or lack of it in a mechanically sensible way. The compaction crossover, "normal" compaction, and "over-pressured" retracement are both related to average effective stress. Figure-1.2 explains both figure-2.6 compaction features as a simple mechanical system that is dependent on mineral physical properties.

Chapter-2.9 Ternary Diagram Loading limb compaction visualization

Figure-2.7 shows the three most common mineralogic end-members as a stacked ternary diagram. The vertical axis of this diagram is effective stress on a logarithmic scale. The logarithmic scale linearizes the power-law effective stress/strain relationship on the vertical scale. The horizontal quartz-calcite-clay plane is linear. Iso-porosity lines are drawn on the surface of the two visible bi-mineral surfaces shown. The limestone-claystone continuum is on the left face of the diagram and the quartz grainstone-claystone continuum is on the right.

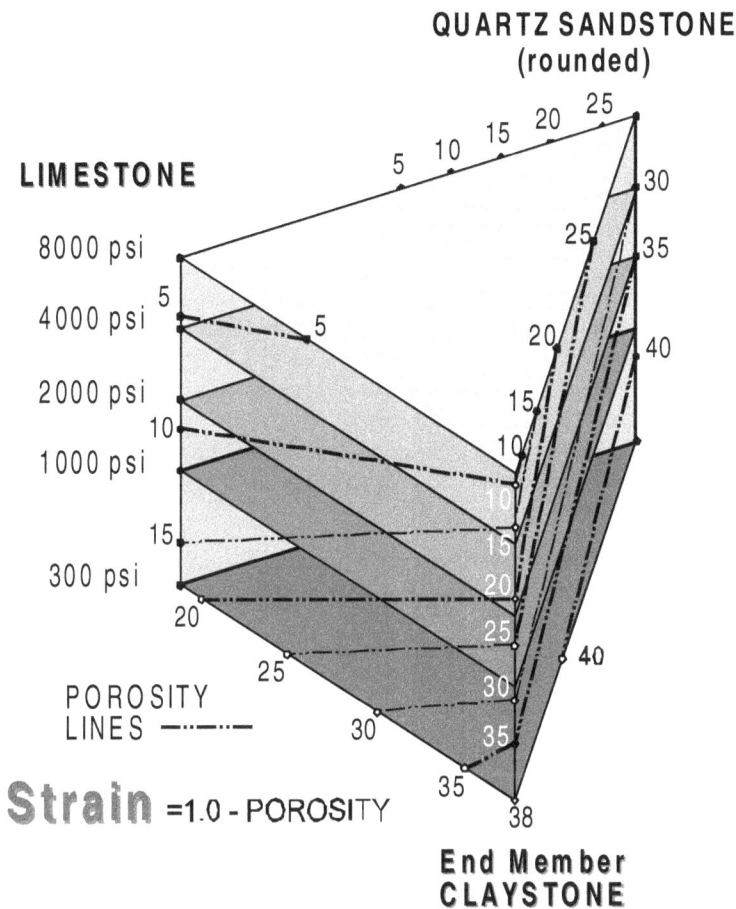

Figure-2.7. Ternary limestone-claystone-quartz grainstone mineralogic compaction resistance diagram. The vertical axis of this diagram is effective stress on a logarithmic scale. Iso-porosity lines are drawn on the surface of the two lithostratigraphic continua shown. The limestone-claystone continuum is on the left face of the diagram and the quartz grainstone-claystone continuum is on the right. This diagram is a mineralogic adaptation of the grain-matrix-compactional stress/strain relationships shown on figure-1.2.

21

The data supporting this diagram is about 300 continuous petrophysical logs from Normal Fault Regime Basins worldwide (Holbrook, 1996). *In situ* measured porosity, mineralogy and effective stress were calibrated at each location. Porosity was estimated from an appropriately transformed density or resistivity log. Lithostratigraphic sequence type, limestone-claystone (L) or quartz grainstone-claystone (S) were assigned by an operator using local knowledge. Relative clay fraction (0 to 1.0) was assigned based upon a baseline normalized gamma-ray log. Holbrook (1995b) describes these procedures.

The features and relative compactional relationships shown on figure-2.7 were observed on all 300 petrophysical logs and are reasoned to be global in nature. The limestone-claystone face (L) of figure-2.7 is characterized by a high-gamma-ray↔high-porosity relationship. The quartz grainstone-claystone face (S) of figure-2.7 is characterized by a high-gamma-ray↔low-porosity relationship. These two lithostratigraphic sequence types produce parallel and hourglass patterns on petrophysical log suites as will be shown on figure-2.8.

Chapter-2.10 The compaction crossover in the quartz grainstone–claystone lithostratigraphic sequence.

The iso-porosity lines on figure-2.7 are traces of iso-porosity surfaces that pass through the ternary mixed mineralogy solid. This three mineral idealization is reasonably close to the gross mineralogic composition of many sedimentary rocks. Many of the patterns that are observed on petrophysical logs are explained by these mineralogic power-law grain-matrix-compactional relationships.

The quartz grainstone-claystone compaction crossover occurs at about 300–500 psi on the right face of figure-2.7. The 35% iso-porosity line is parallel to effective stress on the quartz grainstone – claystone surface. This parallel stress/strain relationship is the physically representative equivalent of a compaction crossover.

Claystones compact more than quartz grainstones at effective stresses above 500 psi. This corresponds to the porosity divergence observed on the figure-2.6 depth function. Any occurrence of calcite in a quartz grainstone – claystone lithostratigraphic sequence will tend to reduce rock porosity. Calcite has significantly lower load-bearing capacity than either quartz or sedimentary clays.

Chapter-2.11 Granular quartz – calcite mineralogic mixtures

Thin isolated limestones often occur in dominantly quartz grainstone – claystone stratigraphic sequences. Where these thin limestones occur, they invariably have lower porosity than the neighboring quartz grainstones. This is because calcite (hardness 3) has much lower compaction resistance than quartz (hardness 7). The calcite–quartz mineralogic continuum is on the back face of figure-2.7 and the expected porosities are shown along the 8000 psi edge of the diagram.

Calcite cement in quartz grainstones invariably reduces porosity. This too is a natural consequence of the softer calcite's low compaction resistance. Whether the calcite mineral is a secondary deposit or not, the softer calcite would yield to the much harder quartz upon loading. Should a calcite grain be re-crystallized or be transported from high to low stress locations by pressure solution, calcite would appear to be cement. The mechanical properties of the minerals present should be considered with the grain-cement nomenclature often used to describe relative compaction.

The calcite mineral lattice (figure-2.1) has the same load bearing capacity whether it is recognized as grain or cement. The same is true of quartz. Minerals bear the average effective stress load through their crystalline lattice irrespective of their geometry. The whole effective stress load is borne by the solid portion of the whole rock that is composed of these minerals.

Chapter-2.12 Limestone – claystone lithostratigraphic sequences

Calcite grainstones, chalk and marl are rock types in the limestone-claystone lithostratigraphic continuum. Porosity is positively correlated to clay content in these sequences. Clay minerals are slightly harder than calcite that increases their load bearing capacity. However, clay mineral inter-particle repulsion accounts for most of the increased porosity with respect to calcite. Much of the water in the marl porespace is probably electrostatically bound as shown on Figure-2.2 c. Pore pressure in limestone-claystone lithostratigraphic sequences can be significant. Loads borne by fluids are the other physical control of sedimentary rock compaction.

Chapter-2.13 Log example of lithostratigraphic controls over the compaction of sedimentary rocks.

SPE 026791   P. HOLBROOK, D. MAGGIORI, R. HENSLEY   11

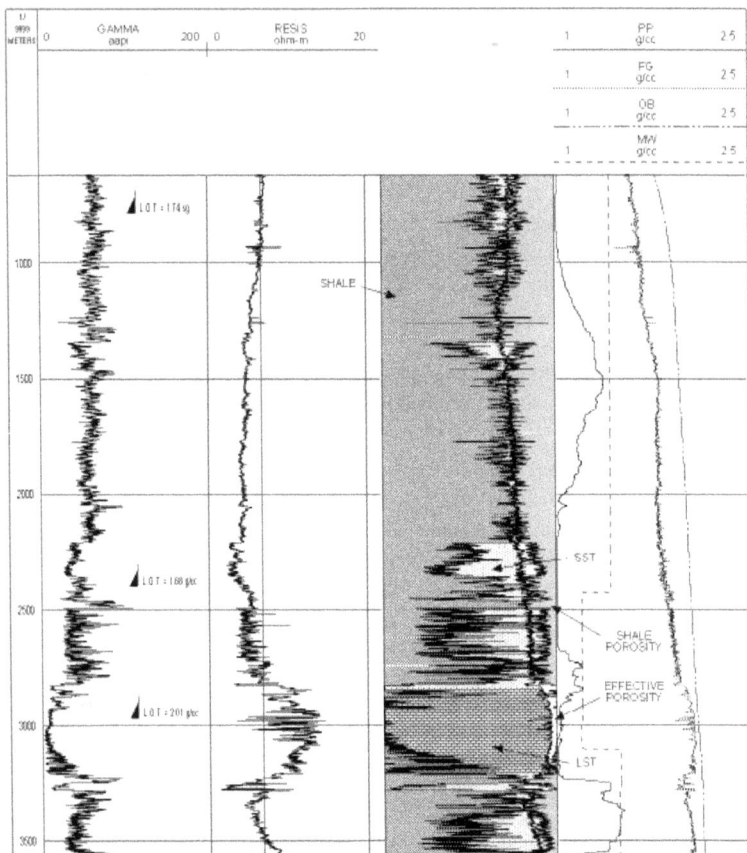

Figure-2.8 is a composite petrophysical log that shows both lithostratigraphic sequence types and their representative log patterns. Track 3 is a lithologic column showing fractional clay volume. Quartz grainstone volume is white. Calcite volume is labeled "lst". Porosity indicated as either bound water (gray shade) or free water (white) is the remainder of whole rock volume in track 3. Figure-2.8 was taken from (Holbrook, P.W., 1995b).

The raw gamma-ray and raw resistivity logs in tracks 1 and 2 move parallel in quartz grainstone-claystone lithostratigraphic sequences. They move opposite in an hourglass pattern in limestone-claystone lithostratigraphic sequences. This is the individual bed scale manifestation of the general power-law compactional relationship portrayed on figures-1.2 and figure-2.7. Below the compaction crossover, quartz grainstones are more compaction resistant than claystones and calcite is less compaction resistant than quartz and claystones. The parallel and hourglass log patterns observed on most logs are related to these (figure-1.2 and figure-2.7) mineralogic compaction resistance relationships.

Chapter-2.14 Inter- and intra-particle Compaction Conclusions

The compaction of sedimentary particles is explained in terms of mineral particle physical properties. The intra-particle load is borne within the mineral's lattice and across electrostatically neutral mineral grain contacts. Compaction of electrostatically neutral particles occurs at each of the "n" contacts with other particles. A power-law stress/strain relationship captures bulk compaction of "n" coordinated particulate solids.

The repulsive electrostatic field between negatively charged clay mineral surfaces is also load bearing. The magnitude of this repulsive field is a power-law relationship to clay particle surface area within a rock volume (Figure-2.3). Inter-particle repulsive force is also a power-law relationship of distance between clay mineral surfaces (Figure-2.2 c). The sum of these clay mineral power-law functions and the electrostatically neutral power-law function is a composite power-law relationship.

The net effects of inter- and intra- particle load types on volumetric in situ strain $(1.0 - \phi)$ are captured with two power-law compactional stress/strain coefficients (σ_{max} & α). The power-law compaction coefficients for the five most common sedimentary minerals were measured from in situ strain after properly accounting for effective stress. Peak granular solid compaction resistance (σ_{max}) is positively correlated to mineral hardness and negatively correlated to mineral solubility. All three are power-law relationships.

End-member sedimentary claystones have a significantly lower (α) than the electrostatically neutral minerals. The unusually high 40– 95% initial porosity of clay rich sediments is power-law related to the log of average sediment particle size. Whole rock compaction is the volume-weighted average (σ_{max}& α) of its individual mineral specific stress/strain coefficients.

This general mineralogic (σ_{max}& α) sedimentary rock stress/strain compactional relationship has been tested in over 300 wells in Normal Fault Regime basins worldwide. Large-scale compaction trends (figure-1.2 & figure-2.6) are explained by this force balanced stress/strain relationship. Observed sedimentary bed scale compactional differences (figure-2.4, figure-2.7, & figure-2.8) are also explained by the same grain-matrix compactional relationship.

The entire load placed upon a sedimentary rock grain-matrix is borne by inter-particle repulsion, intra-particle resistance, and pore fluid pressure. Intra-particle compaction resistance is related to (σ_{max}) and hardness. Inter-particle repulsion contributes additional compactional resistance to (α) in proportion to clay mineral surface area. Mineral ionic bond strength and clay mineral inter-particle repulsion are believed to be primary controls over sediment compaction.

References cited

Berry, L.G. & B. Mason, 1959, "Mineralogy, concepts,descriptions, determinations", W.H. Freeman and Company, San Francisco and London, 630p.

Bryant, W, R Bennett, & C Katherman, 1980, "Shear strength, porosity, and permeability of Oceanic sediments", pp 1555 - 1660. in Vol. 7, "The Sea, the Oceanic Lithosphere", C Emiliani editor, John Wiley & Sons.

Carmichael, R.S., 1982, "Handbook of Physical Properties of Rocks", CRC Press

Carroll, M M, 1980, "Compaction of Dry or Fluid-filled Porous Materials", Journal of Engineering Mechanics Devision, Proceedings of the American Society of Civil Engineers, Vol. 106, No EM5, Oct 1980 pp969 - 990.

Holbrook, P.W, 1995a, "The relationship between Porosity, Mineralogy and Effective Stress in Granular Sedimentary Rocks," paper AA in SPWLA 36th Annual Logging Symposium, June 26-29, 1995.

Holbrook, P.W, D.A. Maggiori, & Rodney Hensley, 1995b, "Real-time Pore Pressure and Fracture Pressure Determination in All Sedimentary Lithologies," pp 215 - 222, SPE Formation Evaluation, December 1995b

Holbrook, P W, 1996, "The Use of Petrophysical Data for Well Planning, Drilling Safety and Efficiency ", paper X in SPWLA 37th Annual Logging Symposium, June 16-19, 1996.

Holbrook, P W, 1999, "A simple closed-form force↔balanced solution for Pore pressure, Overburden and the principal Effective stresses in the Earth.", Journal of Marine and Petroleum Geology, Vol. 16, pp. 303-319.

Hueckel, T. A., 1992, "Water-mineral interaction in hygromechanics of clays exposed to environmental loads: a mixture-theory approach", Canadian Geotechnical Journal V. 29, pp.1071-1086.

Nagaraj, T.S., and Murthy, B.R.S., 1983, "Rationalization of Skempton's compressibility equation," Geophysique, Vol. 33, #4, pp 433-443.

Revil, A. P.A. Pezard & M. Darot, 1997, "Electrical conductivity, spontaneous potential and ionic diffusion in porous media", in Lovell, M.A. & Harvey, P.K., 1997, Developments in Petrophysics, Geological Society special publication, pp. 253, 275

Shumway, G., 1960, "Sound Speed and Absorption Studies of Marine Sediments by Resonance Method - Part II," Geophysics, June, vol. 25, pp. 659 - 682.

Skempton, A.W., 1970, "The consolidation of clays by gravitational compaction," Quarterly Journal of the Geologic Society of London, vol. 125, pp. 373-411, 22 figures.

Stuart, C.A., 1970, Geopressures, in Proceedings of the Second Symposium on Abnormal Subsurface Pressure, Louisiana State University, Baton Rouge, Louisiana, January 1970: 121p.

Terzaghi, K. Van, 1923, "Die Berchnung der Durchassigkeitziffer des Tones aus dem Verlauf der Hydrodynamischen Spannungscheinungen", Sitzunzsber Akad Wiss. Wein Math Naturwiss, K1.ABTS 2a, pp. 107-122.

Chapter-3. COMPACTION vs. DEPTH CURVES related to *in situ* POWER-LAW GRAIN-MATRIX STRESS/STRAIN FUNCTIONS.

The effective stress load bearing capacity of a sedimentary mineral is its combined mechanical and pressure solution compaction resistance. Using gravitational compaction relationships derived from downhole petrophysical data, one can empirically determine the maximum load bearing capacity (σ_{max}) and grain-matrix compressibility exponent (α) as sedimentary mineral properties.

Baldwin & Butler (1985) used solidity (1.0 - porosity) as their strain parameter to normalize the compaction of clay rich sediments and shales from 15 different basins worldwide. They used a power-law effective stress/solidity relationship that is essentially the same as equation-1.3 to normalize their worldwide gravitational claystone compaction data.

Chapter-3.1 Mineral end-member compaction vs. depth and stress/strain functions

Gravitational porosity vs. depth compaction curves are equivalent to Force↔balanced Mineralogic power-law linear Effective Stress / *in situ* Strain (1. - φ) functions.

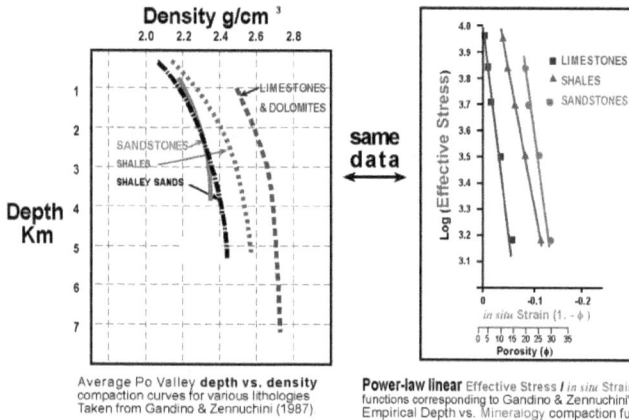

Average Po Valley **depth vs. density** compaction curves for various lithologies Taken from Gandino & Zennuchini (1987)

Power-law linear Effective Stress / *in situ* Strain (1. - φ) functions corresponding to Gandino & Zennuchini's (1987) Empirical Depth vs. Mineralogy compaction functions.

Figure-3.1 shows a set of mineralogic end-member compaction curves measured from petrophysical logs by Gandino & Zennuchini in 1987.

28

The changes in observed bulk density that occur with depth are directly related to porosity, because each mineral has a unique grain-matrix bulk density. These natural geologic loading depth vs. compaction functions are curved because depth is not linear with respect to stress, and porosity is not linear with respect to strain. The curves are widely spread indicating that different end-member sedimentary minerals have very different compaction resistances.

Figure-3.1 right panel shows the same Gandino & Zennuchini compaction data recast as power-law effective stress/strain relationships. Effective vertical stress was calculated by integrating the average bulk density data and subtracting a hydrostatic fluid pressure gradient. Solidity at each kilometer of burial depth was calculated using the appropriate mineral grain density and normal salinity fluid density. The effective stress / grain-matrix-compactional strain relationships for all three single mineral end-member lithologies are essentially power-law linear.

Chapter-3.2 Mechanical unifying concept for basin variable compaction functions.

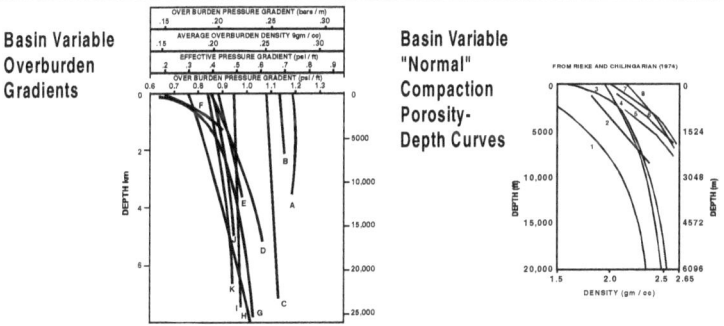

Mineralogic Compaction Effective Stress Unifying Concept

Basin Variable Overburden Gradients

Basin Variable "Normal" Compaction Porosity-Depth Curves

FROM RIEKE AND CHILINGARIAN (1974)

The variation between porosity - depth curves in different basins are all explainable as mineralogic stress / strain functions with basin variable overburden stress (S) and pore pressure (Pp) gradients.

Reike and Chilingarian (1974) compiled the bulk density vs. depth functions from several previous compaction studies. These basin specific claystone compaction curves are number coded on the right panel of figure-3.3. In general the curves go from youngest to oldest sedimentary basins.

The left panel of figure-3.3 is from another compilation of overburden gradient vs. depth functions that are letter coded from highest to lowest overburden gradients. Curves a,b, and c are from thrust fault basins. Their greater density and consolidation is due to the additional horizontal effective stress in thrust fault tectonic regimes.

For the remaining compaction curves there is a correspondence between higher overburden gradient and lower porosity vs. depth for the different basins. Effective stress was not calculated in these studies either. Had these very different compaction vs. depth curves been re-cast as power-law effective stress vs. strain functions, they would probably reduce to a single claystone stress/strain function as was done by Baldwin & Butler in 1982. Mineral specific effective stress/strain relationships (figure-1.2 & table-1.1) are a unifying mechanical concept for all these previous compaction vs. depth studies.

Chapter-3.3 Compaction of Mixed Quartz - Clay Sedimentary Rocks.

The vertical axis on figure-2.7 ternary mineralogic diagram is power-law linear. The iso-porosity lines that bound the triangular solid are almost linear on this binary mineral grain-matrix-compactional surface. The data from several field and area specific studies support power-law linearity of mixed-mineral sedimentary rocks.

Thomas & Steiber (1975) found a linear relationship between shale content and porosity from petrophysical measurements of naturally sedimented and compacted quartz sand - clay mixtures. Figure-3.4 shows the relationship between porosity and shale volume derived from downhole petrophysical data from a 7800 foot depth reservoir offshore Louisiana. The data at constant effective stress are closely scattered about the linear porosity - mineralogy regression line. This supports a linear mineralogic mixing law.

Pittman & Larese (1991) found a near linear relationship between percent ductile grains and porosity for laboratory compacted mixtures. Holbrook (unpublished data) also found a linear relationship between clay content and porosity at several different levels of effective stress in quartz sand – claystone lithostratigraphic sequences. In these cases and the earlier findings of Terzaghi and Skempton, higher ductile grain and clay content resulted in uniformly higher grain-matrix-framework compressibility.

In all four examples cited, the apparent <u>porosity</u> - <u>clay</u> mineral volume mixing-law relationship is essentially linear.

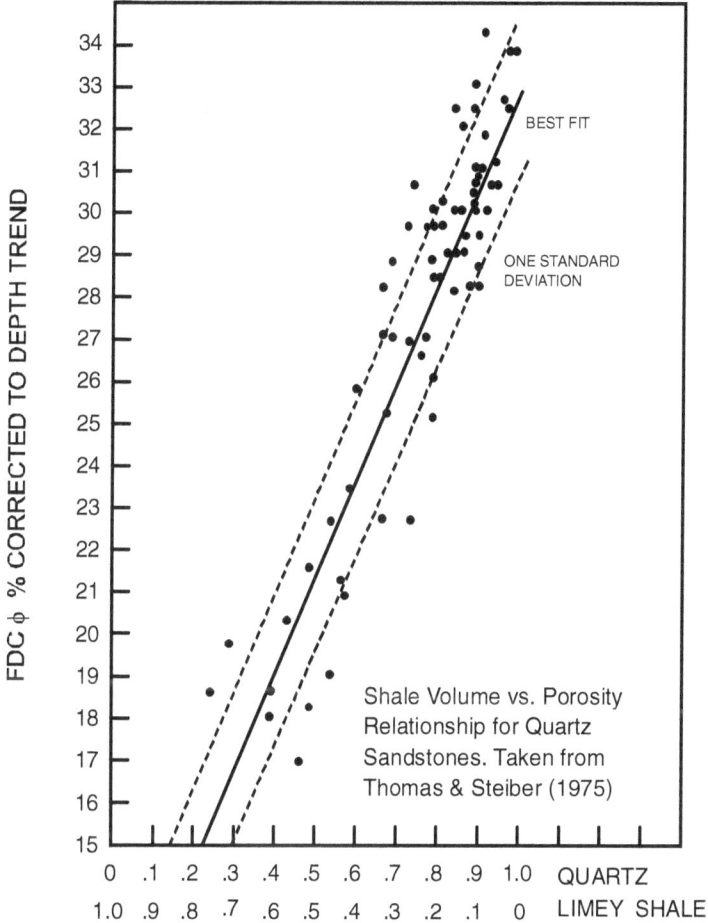

Shale Volume vs. Porosity Relationship for Quartz Sandstones. Taken from Thomas & Steiber (1975)

Chapter-3.4 Compaction of Calcite-Clay Sedimentary Rocks

Jankowsky (1970) found a set of linear compactional velocity relationships in calcitic and shaley sediments in the Upper Cretaceous of northwest Germany. At all levels of effective stress, mixed mineralogy rock velocity and porosity were essentially a linear function of shale volume. Figure-3.5 shown below are Jankowsky's velocity trends.

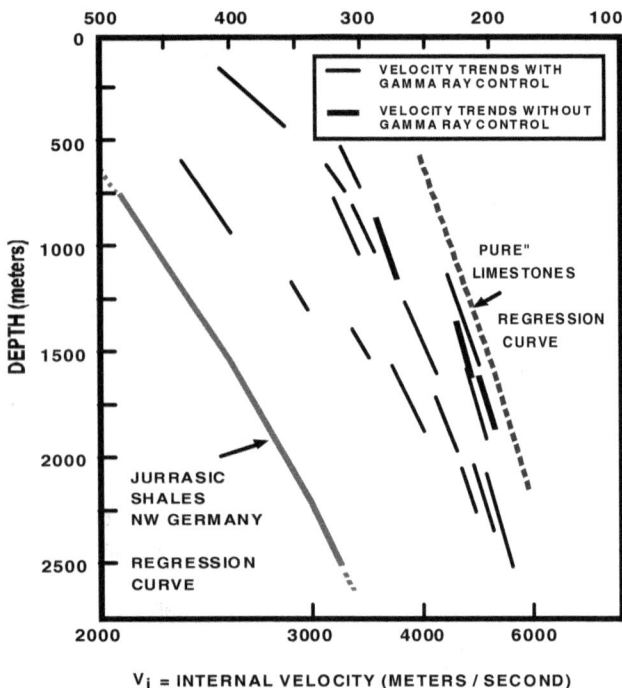

Clay - Calcite *in situ* compactional velocity vs. normalized gamma ray vs. depth relationships from West Germany Chalk Sequences. Taken from JANKOWSKI (1974) Shale volume vs. velocity relationships are quasi linear mineralogy vs. depth functions.

The individual compaction trends on Jankowski's figure-3.5 are sub-parallel to the pure limestone and Jurassic shales compactional relationships. Natural gamma-ray is a good surrogate for relative clay volume in these mixed calcite-clay sedimentary rocks. This data suggests, but does not prove that the mineralogic mixing laws between calcite and clay could be linear.

Holbrook et al. (1993) found a linear relationship between porosity and shale content in Cretaceous limestone - shale stratigraphic sequences in the North Sea. The porosity-mineralogy relationship was linear between the shale and calcite mineralogic end members shown in table-1.1 and figure-1.2. The more soluble limestones and chalks had uniformly lower porosities than marl or shale upon compaction. These consistent observations lead to the conclusion that compaction of calcite - shale mixtures are also an approximate linear function of mineralogy.

Chapter-3.5 Compaction of Granular Quartz - Calcite Sedimentary Rocks

Grigsby & Kerr (1993) found a strong negative correlation between percent calcite and porosity in Type 1 reservoirs in South Texas. Figure-3.6 shows their data derived from thin section microscopic point counts. The quartz-calcite porosity data are more widely scattered. But the data shows the same relative porosity - mineral compaction relationships. Pure quartz sands have higher porosity because of the mineral's higher compaction resistance.

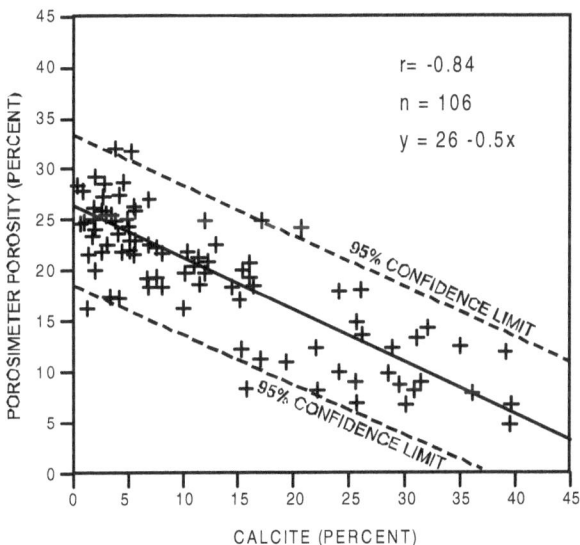

Figure-3.6 The relationship between porosity and mineralogy for quartz - calcite binary mixtures. Taken from Griggsby & Kerr (1993). Solidity $(1.-\phi)$ is approximately equal to that predicted from the average(σ_{max} & α) of the mixed mineralogy sedimentary rock.

Figure-3.6 shows that porosity decreases significantly with increasing percent calcite. This should be expected considering table-1.1 because the maximum compaction resistance (σ_{max}) of calcite is less than 1/10 that of quartz.

Pressure solution is the dominant compaction mechanism for mixtures of these two minerals. Calcite occupies quartz intergranular from mechanical considerations. Quartz is about 1000x harder than calcite and 200x less soluble as indicated in table-1.1. Pressure solution compaction occurs because grain contact pressure raises mineral solubility. If calcite and quartz are in contact under geologic loading, the softer more soluble calcite will be preferentially dissolved by pressure solution at quartz intergranular contact sites. The ions dissolved by pressure solution compaction are usually re-precipitated locally in the quartz intergranular space as calcite cement.

Calcite cement is also precipitated by chemical disequilibrium occurring after long distance fluid migration. Long distance fluid migration or near surface seawater evaporation can result in additional porosity occlusion. Cement resulting from pressure solution compaction may be indistinguishable from this fluid chemical disequilibrium diagenetic calcite cement. There were some megascopically obvious calcite chemical precipitates in the figure-3.6 dataset. Even including these obvious chemical cements, the dominant factor affecting average rock porosity is the mechanical compaction resistance of the constituent minerals.

Chapter-3.6 General Mineralogic Effective Stress Compactional Mixing Law

The first mechanically sensible step is to use the effective stress theorem (Equation-1.2) to calculate the load applied to the rock grain matrix framework. The second step is to relate average effective stress to grain matrix framework compactional strain (solidity) which is done with equation-1.3.

The single mineral compaction coefficients sigma max (σ_{max}) and (α) are empirically derived from *in situ* gravitationally compacted sedimentary rocks. Table-1.1 lists these compaction coefficients as well as other compaction related mineral physical and chemical properties. The binary mixed mineral crossplots indicate that a linear mineralogic mixing law is probably appropriate between the single mineral compaction functions. All the above are resolved with a single generally applicable quantitative relationship.

The mixed mineralogy rock compactional power-law relationship is a mineralogic volume weighted average of the single mineral compactional relationships. To average mineralogic power law linear functions, one must average the power-law slopes (α 's) and sigma max intercepts (log(σ_{max} 's) separately. This mineralogic weighted averaging can be done with a three-step procedure;

1. Calculate the mineralogic volume-weighted-average, of log(σ_{max}) exponents of pure minerals shown on table-1.1.
2. Calculate (σ_{max}) for that mixed mineralogy rock by raising the mantissa to the average log(σ_{max}) exponent.
3. Calculate α for the mixed mineralogy rock as the weighted average of the individual end-member mineral α 's.

Using the mineralogic weighted average σ_{max} and α, porosity and its complement solidity are approximately linear function of mineralogy at all levels of effective stress. Each mineral and all natural sedimentary mineral mixtures are represented by the same (Equation-1.3) power-law effective stress/strain relationship. This combination of the effective stress theorem Equation-1.2 and Equation-1.3 mathematically represents a general mineralogic compactional relationship for nearly all sedimentary rocks. Following this method and approach, gravitational compaction is explained mechanically in terms of sedimentary rock physical properties and the stress applied to the grain-matrix-framework.

Chapter-3.7 Effective Stress Equilibration Time

Effective stress equilibration is virtually certain for compacting subsurface sedimentary rocks. Schatz & Carroll (1981) measured the time dependent creep compaction of porous quartz sandstones under laboratory simulated in situ stress conditions. Linear extrapolation of average observed laboratory creep compaction data indicate that total porosity removal would occur within about 5 years. Table-1.1 indicates that quartz is the most compaction resistant natural sedimentary mineral. At the same effective stress, the less compaction resistant minerals would compact faster. They would reach compaction equilibrium with any level of effective stress (overburden - pore pressure) in less than 5 years.

Dudley et al (1994) also measured creep compaction and developed a power-law relationship to scale between laboratory and reservoir production compaction rates. Their results indicate that equilibration times are less than a few decades.

Sediments below 5 meters depth are usually over 1000 years old. Geologic age increases with depth so that most subsurface sediments have been close to their present depth and effective stress for thousands to millions of years.

Even in the most rapidly subsiding basins, subsidence and sediment accumulation rates are usually less than 5 mm per year. On average one foot of subsidence takes over 200 years. The change in effective stress during 200 years is about 1/2 psi if fluid escape is unrestricted. The change in effective stress is less for undercompacted sediments where fluid escape is restricted.

Comparing laboratory and natural gravitational sediment compaction, the change in effective stress to be accommodated during compaction is very small and the geologic time available for equilibration is very long. Effective stress is immeasurably close to the static force balance of equation-1.2. for hydrostatically compacted and undercompacted sedimentary rocks. Pore pressure resulting from disequilibrium compaction should be calculated using equation-.

Chapter-3.8 Effective stress compactional mixing-law conclusions

The compaction of sedimentary rocks is principally controlled by; the average effective stress load that is applied to the grain-matrix framework, and average grain-matrix mineralogic composition. Stated differently, there is a volumetric stress/strain relationship of the sedimentary rock grain matrix framework.

As proof of this, gravitational compaction of binary mixtures of the three most common sedimentary minerals have linear porosity-mineralogy relationships at all levels of effective stress. A linear mineralogic compaction mixing law is supported by examination of all the mineralogic-porosity data shown on figure-3.3, figure-3.4, and figure-3.5 crossplot data in Chapter-3. These crossplots and additional data document the log-linear effective stress iso-porosity lines portrayed on the faces of the ternary mineralogic diagram figure-2.7.

Linear mineralogic volume-weighted-averaging is accomplished by averaging the logarithms of the individual mineral sigma max (σ_{max}) coefficients; combined with linear volume weighted averaging of the individual mineral power-law grain-matrix compaction (α) coefficients.

The mineralogic weighted averages of sigma max (σ_{max}) and (α) coefficients in equation-1.3 can be used to calculate the porosity of mixed mineral sedimentary rocks from average effective stress and average mineralogy.

References

Atwater, G. L. & E. E. Miller, 1965, "The Effect of Decrease in Porosity With Depth on Future Development of Oil and Gas Reserves in South Louisiana"[abs.]: AAPG Bulletin v. 49, p.334

Baldwin, B. & C O Butler, 1985, "Compaction Curves", AAPG Bulletin, Vol. 69, No. 4, pp. 622-626.

Casas, Enrique & Tim K. Lowenstein, 1989, "Diagenesis of Saline Pan Halite: Comparison of petrographic features of modern, Quaternary, and Permean Halites", Journal of Sedimentary Petrology, (USA) 59(5), pp. 724-739.

Dudley, John W, M T Meyers, R D Shew, M A Arasteh, 1994, "Measuring compaction and compressibilities in unconsolidated reservoir materials via time scaling creep", SPE/ISRM 28026, Rock Mechanics in Petroleum Engineering, August 1994, Delft, The Netherlands pp. 45-54.

Gandino, A. & G. Zenucchini, 1987, "Density Depth Correlation in Po Valley Sediments." Bollettino de Geofisica Teorica ed Applicata, Vol XXIX, pp. 221-231.

Grigsby, J.D. & D.R. Kerr, 1993, Gas Reservoir Quality Variations and Implications for Resource Developent, Frio Formation, South Texas: Examples from Seeligson and Stratton Fields. Texas Bureau of Economic Geology Geologic Circular 93-2 27pp.

Holbrook, Phil, D A Maggiori, R Hensley, 1993, "Real-time Pore Pressure and Fracture Gradient Evaluation in All Sedimentary Lithologies", SPE 26791

Hurley, Michael T, & P Hemphel, 1990, "Porosity and velocity vs. depth and effective stress in carbonate sediments", Proceedings of the Ocean Drilling Program, Scientific Results, Vol 115, pp. 773-777.

Jankowsky, W.1970, , "Empirical Investigation of Some Factors Affecting Elastic Wave Velocities in Carbonate Rocks." Geophys. Prosp. (Netherlands), V. 18, No. 1, pp. 103-118

Pfiefle, T.W. & P.E. Senseny, 1981, Elastic-Plastic Deformation of Anhydrite and Polyhalite as Determined from Quasi-Static Triaxial Compression Tests." Sandia National Laboratories Report SAND81-7063.

Pittman, E.D. & R.E. Larese, 1991, "Compaction of Lithic Sands: Experimental Results and Applications." AAPG Bull V 75, No. 8, pp. 1279-1299, Aug.

Schatz, John T & M M Carroll, 1981, Creep compaction of porous rock; Proceedings of the International Symposium on Weak Rock / Tokyo / 21-24 September 1981, pp. 131-136.

Skempton, A. W., 1970, The consolidation of clays by gravitational compaction, Quarterly Journal of the Geologic Society of London; vol 125, pp 373-411, 22 figures.

Stuart, C. A., 1970, Geopressures, in Proceedings of the Second Symposium on Abnormal Subsurface Pressure, Louisiana State University, Baton Rouge, Louisiana, January 1970: 121p.

Terzaghi, K., 1923, Die Beziehungen zwischen Elastizitat und Innerdruck. Sitzungsbur Akad. Wiss. Wein. 132 Abt. IIa p.205.

Thomas, E.C. & S.J. Stieber, 1975, "The Distribution of Shale in Sandstones and Its Effect Upon Porosity." 16th Annual SPWLA Logging Symp Trans. paper T, 15pp.

Chapter-4.0 STRESS RATIO and STRESS / STRAIN RELATIONSHIPS in NORMAL FAULT REGIME and STRIKE-SLIP BASINS.

In Normal Fault Regime ≈biaxial basins the minimum principal stress is also directly related to *in situ* strain. If pore pressure is known or inferred, the *in situ* minimum principal stress (σ_h) can be measured indirectly from fracture propagation pressure (P_f) using force balance definition Equation-4.1

$$\sigma_h = P_f - P_p \qquad \text{Equation-4.1}$$

An ordinary leakoff test measures fracture propagation pressure (P_f) if there is a sub- vertical fracture in the open borehole interval. An extended leakoff test as described by Kuntze & Steiger (1992) will measure fracture propagation pressure even if the open borehole is initially unfractured.

Many authors have conducted many studies in different basins reporting the variability of fracture propagation pressure (P_f) and minimum principal stress (σ_h) as a function of increasing depth. These empirical depth relationships are helpful but do not explain the underlying physics and are not force balanced. Pilkington (1978) summarized the most widely known relationships at that time. The left panel of figure-4.1a shows his summary.

Figure-4.1a is a typical empirical depth vs. horizontal / vertical stress ratio (σ_h / σ_v) plot. Horizontal / vertical stress ratio (σ_h / σ_v) increases to the right. Both horizontal and vertical effective stresses increase with depth. All the basin specific depth functions on figure-4.1a curve in a similar non-linear manner. Overburden gradient increases with depth in each of these basins accounting for most of the non-linearity in the plot.

The empirical hand drawn curves reveal nothing about the underlying constitutive force balance physic and can only be applied as average depth vs. stress ratio look up tables. The effective horizontal/vertical stress ratio (σ_h / σ_v) on figure-4.1a increases systematically in each of these strictly empirical leakoff test studies from about 0.3 near the surface to about 0.9 at 16000 feet.

39

The Second Fundamental *in situ* Stress Ratio/Strain Relationship

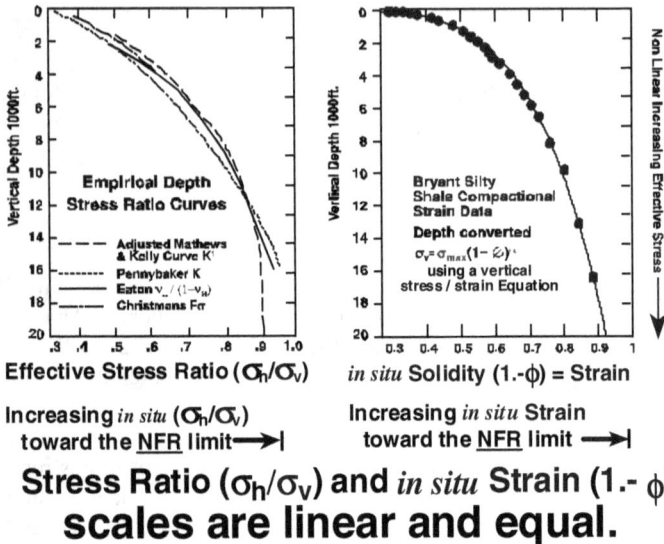

Left panel:
Vertical Depth 1000ft. (0 to 20)

Empirical Depth Stress Ratio Curves

- – – Adjusted Mathews & Kelly Curve K'
- .... Pennybaker K
- —— Eaton v_ /(1–v_H)
- –·– Christmans Fσ

Effective Stress Ratio (σ_h/σ_v)
.3 .4 .5 .6 .7 .8 .9 1.0

Increasing *in situ* (σ_h/σ_v) toward the <u>NFR</u> limit →|

Right panel:
Vertical Depth 1000ft. (0 to 20)

Bryant Silty Shale Compactional Strain Data

Depth converted
$\sigma_v = \sigma_{max}(1-\phi)^c$
using a vertical stress / strain Equation

in situ Solidity (1.–φ) = Strain
0.3 0.4 0.5 0.6 0.7 0.8 0.9 1

Non Linear Increasing Effective Stress →

Increasing *in situ* Strain toward the <u>NFR</u> limit →|

Stress Ratio (σ_h/σ_v) and *in situ* Strain (1.- φ) scales are linear and equal.

Figure-4.1b is the right panel of the above figure. It shows a <u>depth</u> vs. <u>solidity</u> (<u>strain</u>) plot of some measured *in situ* data for silty shales made by Bryant et al (1980). The *in situ* <u>depth</u> data was converted to <u>effective stress</u> using Bryant's bulk density data measured on the same samples. The <u>depth</u> vs. <u>strain</u> (<u>solidity</u>) dataset shown on <u>figure-4.1b</u> was fit with a <u>power-law linear effective stress/strain relationship</u>. The curve shown on <u>figure-4.1b</u> is that <u>effective stress</u> function converted back into a <u>depth</u> vs. <u>strain</u> (<u>solidity</u>) function. The <u>4.1b</u> power-law linear stress/strain function provides a physically rational explanation for both the <u>figure-4.1a</u> and <u>4.1b</u> <u>depth</u> curves.

The fit of the force balanced <u>effective stress</u>/strain function to the 26 points of Bryant's <u>solidity</u> (<u>strain</u>) data on <u>figure-4.1b</u> is excellent. Each data point shown is 1.25 <u>porosity</u> units in radius. This direct stress/strain function touches every data point over the entire <u>depth</u> range. The <u>porosity</u> variability is proportional to absolute <u>strain</u> in Bryant's silty shale dataset. Compactional <u>strain</u> (<u>solidity</u>) is power-law linear with respect to <u>effective stress</u> and force balance is properly taken into account.

The depth curves in figure-4.1a and 4.1b are purposely scaled the same on both horizontal and vertical axes. The hand drawn empirical stress ratio (σ_h / σ_v) curve (4.1a) increase in a manner similar to the effective stress/strain function figure-4.1b which is a direct measure of absolute *in situ* strain. The causal connection between the two Figures is approximately stress ratio equals strain.

The Second Fundamental *in situ* Stress / Strain relationship is;

$$\sigma_h \: / \: \sigma_v = (1.0 - \phi)$$ Equation-4.2

This relationship was hypothesized as a force balanced strain-related predictor of fracture pressure in 1983. Hundreds of leakoff test vs. effective stress comparisons in approximately biaxial normal fault regime basins have been made in the intervening time period. Five separate statistical studies of about 30 leakoff tests each had a uniform 0.4 ppg standard deviation prediction error using this very simple function in conjunction with force balance (Holbrook, 1996a).

Perplexingly, this relationship was found to be insensitive to mineralogy and lithology (Holbrook, 1996a). The Second Fundamental in situ Stress/Strain Relationship gives equally accurate predictions of leakoff test pressures in sandstones, limestones and shales for which it was initially calibrated. Additional lithologic factors were tried in an effort to improve statistics of *in situ* strain predicted vs. measured leakoff tests. All these attempts failed so the very simple Second Fundamental *in situ* Stress/Strain relationship has no additional lithologic coefficients. The ultra simplicity of this relationship suggests that the loading limb compaction of all *in situ* sedimentary rocks in normal fault regime basins is essentially plastic.

The Second Fundamental *in situ* Stress / Strain relationship binds Terzaghi's (1923) Effective Stress "Law " to the effective stress theorem in normal fault regime biaxial basins. This h/v stress ratio to strain linkage (Equation-4.1) physically explains the consistent accuracy of Terzaghi's effective stress "law". Horizontal stress increases in direct proportion to vertical stress and strain. This mechanically sensible conclusion supported by a world of data arises from this closed-form force balanced stress/strain relationship in the earth.

Chapter-4.1 Combined first & second fundamental *in situ* stress/strain relationships

Minimum horizontal stress and fracture pressure are related to mineralogy through a combination of the First & Second Fundamental *in situ* Stress/Strain Relationships under force balance in ≈biaxial normal fault regime basins. Vertical (σ_v), horizontal (σ_h), and average (σ_{ave}) effective stresses are related to each other through the effective stress theorem and Terzaghi's (1923) uniaxial force balanced stress field definitions. Terzaghi's uniaxial approximation compaction "law" is very accurate because in a normal fault regime ≈biaxial stress field, all 3 effective stresses increase together proportionally. A measure of vertical (σ_v) effective stress to predict total *in situ* strain (1.0 - φ) and horizontal stress under these conditions is equivalent to a measure of all three principal stresses.

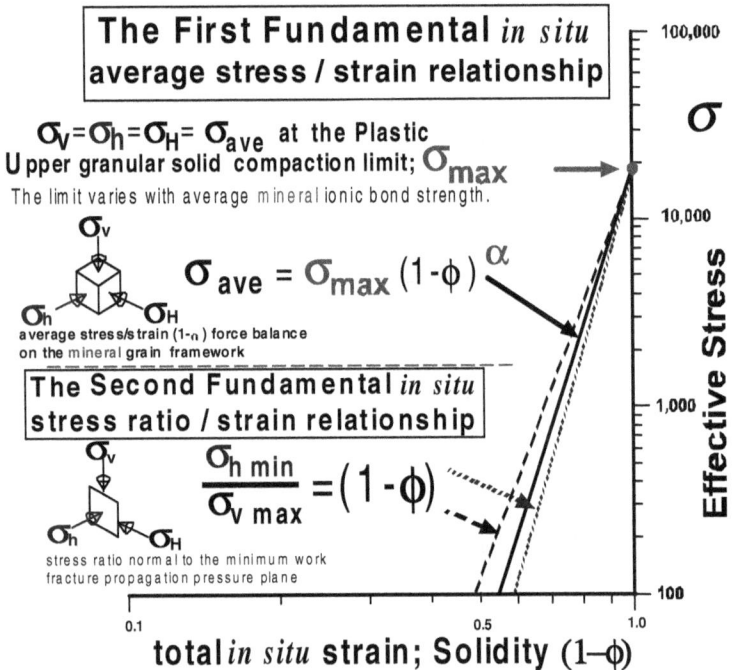

The First Fundamental *in situ* average stress / strain relationship

$\sigma_v = \sigma_h = \sigma_H = \sigma_{ave}$ at the Plastic Upper granular solid compaction limit; σ_{max}
The limit varies with average mineral ionic bond strength.

$$\sigma_{ave} = \sigma_{max} (1-\phi)^{\alpha}$$

average stress/strain (1-α) force balance on the mineral grain framework

The Second Fundamental *in situ* stress ratio / strain relationship

$$\frac{\sigma_{h\,min}}{\sigma_{v\,max}} = (1-\phi)$$

stress ratio normal to the minimum work fracture propagation pressure plane

total *in situ* strain; Solidity (1–φ)

Figure-4.2 shows a shale example of the three force balance definition linked power-law linear effective stress/strain relationship. The

42

compactional effective stress exponent (α) is proportionally different for the three stress/strain relationships (σ_v), (σ_h), and (σ_{ave}) shown on figure-4.2. Vertical effective stress is the maximum in Normal Fault regime basins. The subscripts; v = **vertical**, h = **minimum horizontal**, and ave = **average**, apply to the effective stresses shown on figure-4.2. The vertical axis is effective stress of the sub-scripted variable and the horizontal axis is total *in situ* Strain ie. Solidity.

All three stress/strain functions intersect at the same (σ_{max}) total compaction intercept where solidity (1.0 - ϕ) equals 1.0. At this total compaction intercept the entire external load is borne by the ionic bonds of the minerals. The effective stress theorem (Equation-1.2) limits are honored in that; (S_{ave}) = (σ_{ave}) and (P_p = 0.0). An ideal plastic at the (σ_{max}) compaction limit would equilibrate to an isotropic stress condition wherein [(σ_v)= (σ_h)= (σ_{ave})].

The total *in situ* grain-matrix-compactional strain is very close to ideal plastic strain throughout the compaction history of a sedimentary rock. This is physically reasonable when one considers that less than 3% of the total *in situ* strain is elastic at the (σ_{max}) total compaction pressure intercept for the common sedimentary minerals (Carmichael, 1982). The reversible thermal expansion of these same minerals is about -3% at average *in situ* temperatures that prevail at (σ_{max}). Thus the two reversible strains; elastic and thermal, almost exactly cancel each other under average *in situ* conditions. The net *in situ* strain that is measured with downhole petrophysical sensors is therefore almost entirely (100% +/- 2%) plastic.

The three principal stresses (σ_v , $\sigma_H = \sigma_h$) average to (σ_{ave}) over the whole data range using the stress/strain functions shown on Figure-4.2. The three principal effective stresses are internally force balanced. The effective stress theorem balances all three with pore pressure at the limits and throughout the entire porous granular solid domain.

The total *in situ* stress/strain relationship is essentially plastic. The net reversible strain limit for *in situ* minerals at (σ_{max}) is very small. Reversible strain ranges from (0%) at the surface to (+/- 2%) at (σ_{max}). Thus, dominantly plastic mineral physical properties combined with the effective stress theorem force very close convergence of (σ_v &σ_h) at (σ_{max}) as indicated on figure-4.2. Under *in situ* loading limb conditions the match is an **exactly closed form** because the elastic strains are incorporated into the total *in situ* strain.

The combined first and second fundamental *in situ* force balanced stress/strain relationships have proven to be very accurate through comparison to measured leakoff tests to calculated fracture pressures in hundreds of wells worldwide (Holbrook, 1996). Greater statistical accuracy is an objective measure of the validity of competing scientific theories. The simple closed form force balanced solution is statistically more accurate than any combination of empirical methods tried on the same data.

Chapter-4.2 Stress ratios and strain in Strike-slip fault bounded basins.

The Strike-slip tectonic fault regime has the relative order of stresses;
Maximum Horizontal (S_H) > Intermediate Vertical (S_v) > minimum horizontal(S).

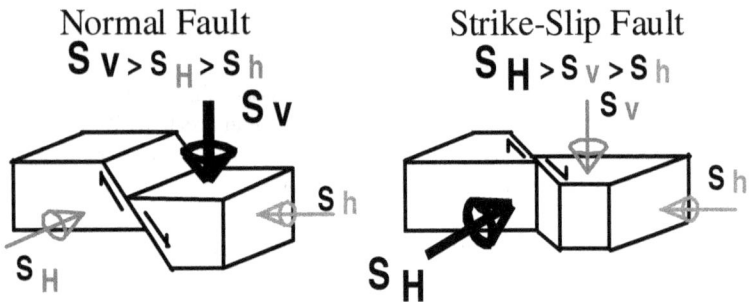

Normal Fault
$S_V > S_H > S_h$

Strike-Slip Fault
$S_H > S_v > S_h$

Shear fractures in the Normal Fault Regime have relative slip directions indicated by arrows on figure-4.3 shown above. When the maximum principal stress is horizontal, shear fractures have a vertical dip and a horizontal displacement indicated by arrows on the right. These are two of three Andersonian fault and basin classifications.

Strain in each tectonic regime is accomplished by slip along shear fractures and by compaction of the rock grain-matrix. The effective stress theorem ($S_{ave} = \sigma_{ave} + P_p$) force balance boundary condition applies to all tectonic regimes.

Solidity is a measure of absolute volumetric *in situ* strain. In ≈biaxial Normal Fault Regime basins, the h/v stress ratio was found to be directly proportional to absolute *in situ* strain (Equation-4.1). Figure-4.1a and figure-4.2 demonstrate these relationships. Using the biaxial definition, the V:H:h stress ratios are in a +2:-1,-1 ratio with respect to the average

44

effective stress. The forces about the average are balanced from the mudline to complete consolidation by the same (Equation-4.1) relationship.

Equation-4.1 is a force balance idealization for horizontally bedded strata in Normal Fault Regime biaxial basins. This idealization does not apply to shear fractures or strata with over 5 degrees of dip (Holbrook, 1999). Nonetheless, this simple idealization has proven to be very accurate at predicting leakoff tests and fracture propagation pressure (Holbrook, 1997).

Kieth Katahara (1995) found an equally surprising stress ratio relationship in the Long Beach Unit in California. He found that " the mean horizontal effective stress to the vertical effective stress is constant from one formation to another". Katahara et al determined these ratios from leakoff tests, minifrac and microfrac measurements in this strike-slip fault regime basin. Katahara's data is plotted on figure-4.4. This figure has been modified from Katahara's figure 2 to reflect effective stress (σ) rather than confining loads (S).

Effective stress profiles for the Long Beach Unit minimum, vertical and maximum vs. TVD

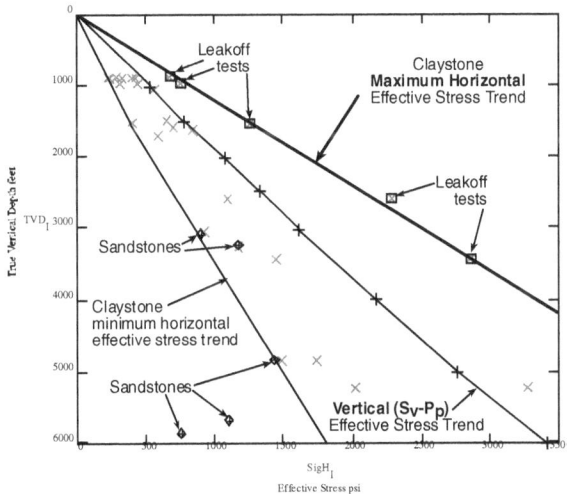

Minimum horizontal, Vertical, and Maximum Horizontal Effective Stresses are in an approximate 1:2:3 Stress Ratio relationship. Adapted from Katahara et al (1995).

There is a slight increase in overburden gradient in the Long Beach unit that imparts a slight curvature to calculated effective vertical stress on figure-4.5. The black cross effective vertical stress values on figure-4.5 were calculated from the porosity log (figure 5) of Katahara et al (1995).

The Long Beach unit produces from the lower Pliocene Repetto Formation and the upper Miocene Puenta Formation. These are both turbidite quartz sand – claystone lithostratigraphic sequences at vertical depth from 2000 to 7000 feet. There are very steep dips and many faults in the Long Beach Unit that would normally complicate the results of the measurements made.

Nonetheless, Katahara et al's data falls within the limits shown and support their previously mentioned constant stress ratio conclusions. An additional observation and conclusion can also be drawn from this data. The effective stresses are in an approximate 1:2:3 ratio with respect to H: v: h limits shown on figure-4.4. The three principal stresses in this even number ratio are a vectorial and volumetric force balance.

The initially surprising commonality between figure-4.2 and figure-4.4 is the apparently even number relationship between the principal effective stress limits. Effective vertical stress has a relative even number value of 2 in both tectonic regimes. There is a difference of + or - 1 from vertical for the horizontal stress limits in both of these Andersonian fault regimes.

Vertical effective stress is relative even number 2 on figure-4.4. Without reliance on numbers, average effective stress would equal the intermediate vertical effective stress. If vertical is both intermediate and average, the greater H and lesser h must differ by an equal amount from the intermediate average principal stress. The 1:2:3 even number relationship on figure-4.4 represents a vectorial within volumetric force balanced relationship for strike-slip tectonic regimes. Combining these three unequal h<v<H integers 1:2:3 for the strike-slip tectonic fault regime produces;

$$\sigma_{ave} = \sigma_v \, , \, \sigma_h = \tfrac{1}{2}\sigma_v \, , \text{ and } \sigma_H = 3/2\,\sigma_v \, ; \quad \text{Equations-4.3}$$

These even number stress ratio relationships are materially no different from the Katahara et al (1995) results for calculating fracture pressures in this strike-slip fault regime oilfield. The primary difference between Katahara's figure 2 and figure-4.4 is the reversal of the depth axis and in accounting for pore pressure. Gravitational force directed toward the center of the earth is the *a priori* reason for the relative stresses to be even numbers. In both the biaxial Normal Fault Regime and the Strike-slip fault regime tectonic regimes, the observed and theoretical match. They are the lowest differing even number combinations of 3 principal stresses that sum to an even number average. These even number combinations match both Andersonian NFR and Strike-slip fault regime relative stress order criterion.

The results from the leakoff tests made in sandstones on figure-4.4 are inconclusive. In two cases they are similar to adjacent claystone leakoff tests. In three other cases they are unusually low. There are too few data to establish a trend. In NFR basins sandstones usually exhibit a trend parallel to but lower than claystones. There are too few data at the Long Beach Unit and probably too many secondary complications to demonstrate this.

Chapter-4.3 Comparison of strain between NFR and Strike-slip fault bounded basins

There are clearly evident grain-matrix-compactional relationships in the Long Beach Unit; figure 5 in Katahara et al's (1995) paper. The gamma-ray log trace was used in conjunction with the porosity to identify end-member sandstones and claystones. A tangent line was drawn through the cleanest quartz sandstones and the highest gamma-ray claystones. These end-member tangent lines are shown in gray on figure-4.5.

The cleanest sandstones have the highest relative porosity and claystones have the lowest porosity as in other sand-shale lithostratigraphic sequences. The quartz sandstone tangent line shown in gray shows a kink at about 1400 psi effective stress. The end-member claystone trend is very uniform with respect to vertical effective stress and TVD.

A linear trend line was drawn through each mineralogic end member group. Solidity increases with increasing vertical effective stress. Grain-matrix compaction is volumetric and affected proportionally by all three principal stresses. The vertical effective stress, even though it is not the maximum, exerts a dominant effect over the average effective stress and resultant grain-matrix compaction.

In both Normal Fault Regime and Strike-slip fault regime tectonic settings overburden increases with depth below the surface as the accumulated bulk density. In both settings, the horizontal effective stresses increase in proportion to the vertical. Figure-4.2 shows these relative stress relationships in Normal Fault Regime basins and figure-4.4 shows these relationships in a strike-slip fault regime basin. All three effective stresses are zero at the earth's surface or mudline and increase proportionally below.

Compactional effective stress profiles for end-member sandstones and claystones in the Long Beach Unit, California

Adapted from Katahara et al (1995).

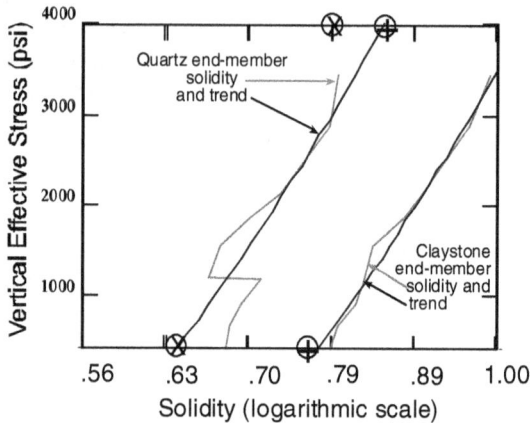

Average Effective Stress and Solidity of end-member quartz
⊗ sandstones and
⊕ claystones
in Normal Fault Regime Basins

Average effective stress grain-matrix compactional profiles in NFR and Strike-Slip tectonic regimes overlap.

Figure-4.5 shows sub-parallel compaction trends for both end-member sandstones and claystones. Quartz and clay resists compaction by about the same amount in either tectonic settings.

For comparison the quartz grainstone and claystone values at 1500 and 4000 psi for Normal Fault Regime basins were transferred from figure-1.2 to figure-4.5. These are the "O" symbols on the upper and lower horizontal borders of figure-4.5. There is substantial overlap between Normal Fault

Regime and Strike-Slip compaction with respect to average effective stress. They coincide fairly closely at 1500 psi, but differ at 4000 psi.

It was not possible to do a detailed petrophysical analysis on the Long Beach Unit data. Compaction may be greater in strike-slip fault regime basins due to secondary non-stress factors. Geothermal gradients are generally higher in strike-slip basins. Water circulation patterns are different in strike-slip fault regime basins due to the very deep strike-slip faults.

Allowing for these unknowns, it is reassuring that grain-matrix compaction is similar in Normal Fault and Strike-slip fault regime tectonic regimes at low levels of average effective stress.

Chapter-4.4 Effective Stress Ratio Conclusions, and the even number tendency

Vertical effective stress is the dominant component of average stress in Normal Fault Regime Basins. Vertical effective stress is the intermediate regulating component of average stress in Strike-Slip Fault Regime Basins. Horizontal effective stresses vary closely with the vertical in both tectonic settings by apparently even number relationships. See figure-4.2 and figure-4.4.

Gravitational stress is the maximum in Normal Fault Regime Basins. If the earth's surface is flat lying, the v:H:h orthogonal coordinate system is oriented about the vertical. Horizontal effective stresses are a compactional response to the greater vertical through grain-matrix strain. It is surprising though supported by measurements that average horizontal stress is directly related to grain-matrix-compactional strain (Equation-4.1) by an apparently even number stress ratio relationship.

The even number tendency probably results from very long time load equilibration within the solid grain matrix (chapter-3.7). Mineral solubility increases at points of higher stress along grain-grain contacts. Ions will tend to dissolve at points of higher stress and re-precipitate in locations of lower stress. This ionic migration both broadens and equalizes stress along all grain contacts. Figure-2.2 a illustrates this slow operating chemical stress equilibration process.

The pressure solution equilibration process buffers unequal effective stress. The vertical effective stress is fixed in the direction toward the earth's center. The two horizontal stresses will tend to equal each other in

Normal Fault Regime basins. The three orthogonal stresses must average to the mean honoring static force conservation. When the gravitational force is maximim, the stress ratios tend toward +2:-1:-1 about the average (0). In Strike-slip fault regime basins, gravitation force is the intermediate principal stress.

When the gravitational force is intermediate, force conservation dictates that the difference in stresses from the average = vertical will tend toward equality. This also tends toward an even number relationship of 1:2:3 of h:v:H about an average of (2). figure-4.4 from the Long Beach field exhibits this even number tendency. It is reasonable to extrapolate that other Strike-slip fault regime basins would also have a 1:2:3 stress ratio relationship where the land surface is nearly horizontal.

In Strike-slip fault regime basins, gravitational stress is a side force to the maximum horizontal stress. Gravitational stress then acts as a 1:2:3 proportional regulator to the maximum horizontal stress. Potentially huge horizontal stress originates along sliding continental plate boundaries. The maximum horizontal stress is transmitted horizontally across the continental margin sub-blocks and relieved vertically to the surface. The magnitude of confining vertical gravitational stress apparently regulates the peak magnitude of the greater horizontal effective stress in porous sediments as shown on figure-4.4. This too is an apparently even number 1:2:3 stress ratio relationship.

There is considerable commonality between grain-matrix compaction in Normal Fault Regime and Strike-slip fault regime tectonic settings. The mineral quartz has much greater compaction resistance than clay. There are mineral specific vertical grain-matrix-compactional profiles in both tectonic settings. At low to moderate effective stress magnitudes and temperatures these compaction profiles are nearly equal volumetric stress/grain-matrix-compactional strain relationships. figure-4.5 and figure-1.2 from different tectonic settings are quantitatively comparable.

Force balance dictates that the three principal stresses must average to the volumetric effective stress. This force conservation boundary condition applies to all tectonic settings at all depths. The maximum, intermediate, minimum and average effective stresses are related to each other through force conservation (average = (1 + 2 + 3)/ 3).

50

The 1:2:3 relationship of H: v: h honors the above mentioned volumetric force balance in strike-slip fault regime basins. If vertical is the intermediate principal stress, force conservation dictates that H and h must differ from vertical by an equal amount. It is encouraging that Katahara's (1995) measurements bear this out. Equations-4.3 are an expression of this force conservation. The 1:2:3 relationship is consistent with force balanced mechanical principals where the earth's surface is relatively flat lying.

The +2:-1:-1 relationship for V:H:h in Normal Fault Regime basins is also force balanced with respect to the volumetric effective stress. The greater H and lesser h would be equal if both the earth's surface and bedding planes are at low dip angles. This $\cong$ biaxial stress field condition exists throughout most Normal Fault Regime basin volumes that are at low dips away from major fault boundaries.

Both Strike-slip fault regime and Normal Fault Regime basins contain relatively large volumes of rock that are close to the mechanical idealizations mentioned above. V:H:h stress fields are rotated in the immediate neighborhood of faults. The 1:2:3 and +2:-1:-1 even number relationships apply to the rock volumes between faults. The data on figure- figure-4.1b, figure-4.2, and figure-4.4 seem to bear this out.

Many other factors are present near and in faults that tend to alter the idealized even number intra-fault relationships. Though the bedding planes and faults of the earth are often geometrically complex, the forces measured in unfaulted rock units are apparently simple mathematical relationships. There is almost certainly gross force conservation across faults as well. While we may not know the orientation of maximum, intermediate, and minimum principal stresses, the average in and around faults should be about the same as the average between faults. Knowledge of the average stress between faults can be applied to solve geologic and engineering problems in and around faults.

Compaction trends are nearly linear because the controlling stress/strain relationships amid the geometric complexity are volumetric. Volumetric averages are insensitive to the orientation and magnitude of the 3 principal stresses. These volumetric stress/strain relationships are strongly apparent whether the vertical gravitational force is the driver or the regulator of average effective stress.

The theory presented in this chapter and book is that earth loads directly affect volumetric grain-matrix-framework compactional strain. The principal v:H:h stresses combine to form the volumetric average. The data in this chapter indicate that there probably is a rock internal relative stress ratio regulating relationship in Normal Fault Regime and Strike-slip fault regime tectonic settings. Though stress fields may be and often are rotated in the neighborhood of faults, force conservation dictates that the three principal stresses must sum to the average.

Many subsurface engineering applications can make use of these average volumetric and relative stress/strain and even number stress ratio relationships. These load, stress, and strain relationship can be easily calculated from conventional petrophysical data. Overburden can be estimated with considerable accuracy from bulk density or equivalent porosity logs. Continuous load, stress and rock properties logs can be generated from the quantitative relationships in this chapter at very low cost. These logs represent related vectorial within volumetric force balance that is apparent and measurable throughout most of the earth's sedimentary crust!

References

Bryant, W., R. Bennett, & C. Katherman, 1980, "Shear strength, porosity, and permeability of Oceanic sediments", pp 1555 - 1660. in Vol. 7, "The Sea, the Oceanic Lithosphere", C Emiliani editor, John Wiley & Sons.

Carmichael, R.S., 1982, "Handbook of Physical Properties of Rocks", CRC Press.

Holbrook, P W, 1999, "A simple closed-form force↔balanced solution for Pore pressure,
Overburden and the principal Effective stresses in the Earth.", Journal of Marine and Petroleum Geology, Vol. 16, pp. 303-319.

Holbrook, P.W., 1995, "The relationship between Porosity, Mineralogy and Effective Stress in Granular Sedimentary Rocks", paper AA in SPWLA 36th Annual Logging Symposium, June 26-29, 1995.

Holbrook, P.W., D.A. Maggiori, & R. Hensley, 1995, "Real-time Pore Pressure and Fracture Pressure Determination in All Sedimentary Lithologies", pp 215 - 222, SPE Formation Evaluation, December 1995

Holbrook, P W, 1997, "Discussion of A New Simple Method to Estimate Fracture Pressure Gradients", SPE Drilling & Completions, March 1997, pp.71-72

Katahara, K. W., K.W. Lynch, and R.G. Keck, 1995, "A Semi-Empirical Model for In-Situ Stress Distribution for a Strike-Slip Regime: The Long Beach Unit, California", SPE 29602, pp 581-592Kuntze, K.R. & R.P. Steiger, 1992, "Accurate In-Site Stress Measurements During Drilling Operations", SPE 24593, pp. 491-499.

Lorenz, J.C., N.R. Warpinski, P.T. Branagan, A.R. Sattler, 1989. " Fracture Characteristics and Reservoir Behavior, in Stress-Sensitive Fracture Systems in Flat-Lying Lenticular Formations", JPT June pp615-622.

Pilkington, P E, 1978, "Fracture Gradient Estimates in Tertiary Basins", Petr. Eng. International, May 1978, pp138-148.

Terzaghi, K. Van, 1923, "Die Berchnung der Durchassigkeitziffer des Tones aus dem Verlauf der Hydrodynamischen Spannungscheinungen", Sitzunzsber Akad Wiss. Wein Math Naturwiss, K1.ABTS 2a, pp. 107-122.

Chapter-5. MINERAL and FLUID SENSITIVE POROSITY TRANSFORMS

The physical properties of the minerals and fluids that compose the earth affect all petrophysical sensors. Each petrophysical sensor responds to a different set of mineral and fluid coefficients depending on governing physics of that sensor. Mineral and fluid properties must be taken into accounted for any petrophysical sensor → porosity transform in order to be accurate.

Chapter-5.1 Factors Affecting Calculation of Porosity from Petrophysical Sensors

The calculation of very accurate porosity (ϕ) from petrophysical sensor data is equivalent to the calculation of very accurate *in situ* strain (1. - ϕ) for further rock mechanics calculations. For calculation purposes there are several mineral coefficients that are essentially constant ($\pm$ 0.04%) under natural *in situ* PV/T conditions. However, clay mineral grain density varies by as much as 16% under natural subsurface conditions.

Most subsurface sedimentary rocks contain Sodium Chloride brine under different PV/T x-Salinity conditions. Under the known range of geothermal gradients in sedimentary basins: 1.) The velocity of Sodium Chloride brine varies by up to 27%; 2.) The density of Sodium Chloride Brine varies from 0.92 to 1.26 g/cc also (27%); and, 3.) The compressibility of Sodium Chloride brines varies by as much as 61% under known *in situ* conditions (Holbrook, Goldberg & Gurevich, 1999b). The electrical conductivity of Sodium Chloride brines (C_w) varies by 5 orders of magnitude. All this natural subsurface variability should be taken into account if one expects to calculate porosity accurately from any petrophysical sensor.

Chapter-5.2 Claystone Diagenetic Effects on Density and Calculated Porosity

During burial diagenesis clay minerals are transformed from low density highly disordered weathering products into well-crystallized metamorphic micas and chlorites. In moderate to high geothermal gradient areas this complete diagenetic transformation occurs in the upper 5000 meters of the Earth's crust. The average grain density of clay minerals in mudstones increases from about 2.64 g/cc to 3.15 g/cc as the minerals are transformed primarily through increasing temperature.

Huang (1991) successfully modeled smectite to illite clay mineral transformation as a thermo-kinetic process in many basins with different geothermal gradients. Aja & Rosenberg (1993) went further showing data supporting thermodynamic equilibrium between clay minerals during this primarily temperature-controlled diagenetic transformation.

Within a given region, the zero porosity average clay mineral grain matrix (ρ_{clay}) density/depth gradient is estimated. Quartz and calcite have essentially constant densities of 2.65 g/cc and 2.71 g/cc in subsurface sedimentary rocks over all known geothermal gradients. Whole rock grain density changes for each sample interval in proportion to the volume-weighted average of clay and non-clay minerals present. When applied this procedure reconciles most of the differences between resistivity sensor and γ–γ density sensor calculated porosities. Accounting for the regional (ρ_{clay})/depth density gradient, results in greatly improved porosity, overburden, effective stress, and pore pressure calculations no matter what petrophysical sensor is used for the primary porosity determination.

Chapter-5.3 Porosity from γ–γ Density Sensor Input

Both average mineral grain density (ρ_{min}) and average downhole fluid density (ρ_{fluid}) must be used to obtain accurate porosity (ϕ) calculations from bulk density (ρ_{bulk}) sensor input. Average clay volume of the solid fraction often varies considerably with depth. A borehole attenuation-corrected natural gamma-ray signal from the formation is used to estimate clay volume (V_{clay}) as a fraction of solid for each foot. Depending upon the stratigraphic sequence type, average mineral grain matrix density is calculated in one of two ways. The average grain matrix density (ρ_{matrix}) used in quartz sand - claystone stratigraphic sequences is;

$$(\rho_{matrix}) = [\,(1.0 - V_{clay}) \bullet 2.65\,] + (V_{clay} \bullet \rho_{clay}) \qquad (5.1)$$

In a calcite - claystone stratigraphic sequence, calcite density (2.71 g/cm^3) is substituted for quartz density (2.65 g/cm^3) in the equation above. The average claystone grain matrix density (ρ_{clay}) increases gradually with depth determined from the regional geothermal (ρ_{clay}/depth) profile.

Fluid density (ρ_{fluid}) is estimated from a regional PV/T x-NaCl salinity profile. The density and bulk modulus coefficients of Sodium Chloride brines were extracted from voluminous measured density and velocity data by Archer (1992). Archer's Equation-Of-State thermodynamic molecular interaction coefficients were re-cast as third order functions of NaCl brine density, pressure, temperature and molality. The third order PV/T x-NaCl regression of NaCl brine Equation-Of-State provides very accurate physically consistent fluid coefficients for porosity (ϕ) from γ–γ bulk density (ρ_{bulk}) measurements estimation, and for Gassmann equation forward and inverse modeling (Holbrook, Goldberg & Gurevich, 1999b). Using the appropriate downhole density coefficients, porosity is calculated from bulk density using the equation:

$$\phi = (\rho_{bulk} - \rho_{matrix}) \, / \, (\rho_{fluid} - \rho_{matrix}) \quad \text{(5.2)}$$

This procedure incorporates the best knowledge we have of grain matrix and fluid densities into the porosity from bulk density calculation.

Chapter-5.4 Porosity from Resistivity Sensor Input

The common sedimentary mineral grains are infinitely resistive to electric current. Sedimentary rocks conduct electricity primarily through the motion of charged Na+ and Cl- ions in the dominantly NaCl brine-filled pore space. The formation resistivity factor (F) relates formation water conductivity (C_w) to measured true rock conductivity (C_t). When porosity (ϕ) = 1.0, F=1.0, and $C_w = C_t$.

$$F = C_w \, / \, C_t \quad \text{(5.3)}$$

Archie (1941) found a power-law relationship between porosity and formation factor for many sedimentary rocks. The combined intergranular tortuosity-cementation exponent (m) varies with porosity and mineral grain shape in the Archie equation;

$$\phi = F^{(-1.0/m)} \quad \text{(5.4)}$$

Most subsurface brines have salinities equal or above normal salinity seawater. At these salinities, almost all the total conductivity (C_t) is through the charged Na+ and Cl- ions in the water phase (C_w). Variability in the tortuosity-cementation exponent (m) is the pore geometric factor in estimating porosity from resistivity. The power law exponent (m) is a measure of the electrical length / actual length of an insulating porous granular solid. It is a complex function of intergranular pore volume, continuity, and shape.

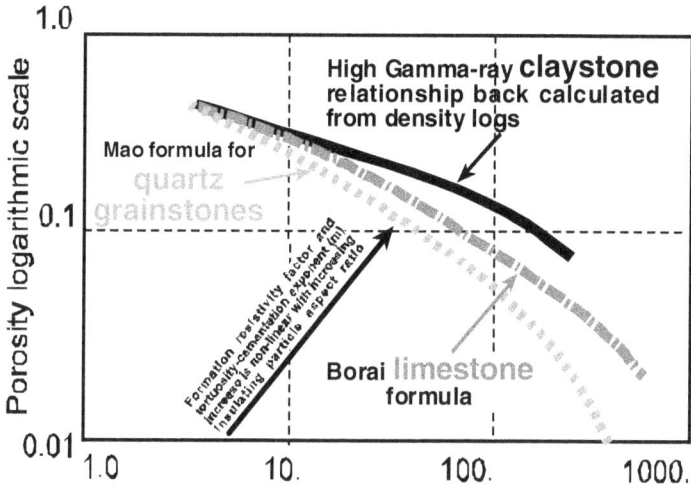

Figure-5.1 shows a set of three measured Formation Factor vs. Porosity relationships for the three most common sedimentary minerals. Implicit in each of these curves is the natural grain shape (aspect ratio) of that insulating mineral as it occurs naturally. Grain aspect ratio varies from (1:1) for perfectly rounded quartz grains to over (500:1) for an average sedimentary clay.

The electrical path though a claystone is many times longer than the electrical path through a quartz grainstone. Taking mineralogically variable tortuosity-cementation (m) into account the formation factor for a 10% porosity grainstone is 50. The formation factor for a 10% porosity claystone is 240. The inappropriate use of a quartz sandstone formation factor to estimate porosity in a claystone would result in (15/10 PU) or 50% over-estimate of claystone porosity.

Mao et al. (1995) determined the quartz grainstone Archie function from laboratory measurements on 155 quartz grainstone core samples. Mao et al's dataset is many times larger than the earlier Archie or Humble datasets. The value of "m" increases non-linearly with decreasing porosity in all three single mineral empirical formation factor relationships.

Borai (1987) developed an empirical formation factor vs. porosity relationship for an equally large core sample dataset for pure limestones in Abu Dhubi. Both "a" and "m" vary in Borai's dataset, but the formation factor trend is offset to higher "m" values than for quartz grainstones. Mineralogically pure limestones are composed of more platy organic particles. Their average aspect ratio is usually higher than the generally equant quartz grains. The increased tortuosity-cementation exponent "m" is evident over the entire range of observed porosities between the rounded quartz and platy limestone datasets (figure-5.1).

Holbrook (unpublished data) calculated end-member claystone formation factors from density and resistivity logs on 5 wells containing only quartz grainstones and claystones. The depth range was from 1000 to 20,000 feet covering a wide range of quartz and claystone porosities. The density logs showed unequivocally that the claystones had much lower porosities than the approximately depth equivalent quartz grainstones. A formation factor ratio ($F_{claystone}$ /$F_{quartz\ grainstone}$) was developed from this dataset. The claystone formation factors were leveraged from the well-determined quartz grainstone formation factors. The result is the uppermost "m" variable claystone formation factor relationship shown on figure-5.1.

All three of the mineralogic end-member formation factor vs. porosity relationships were determined from large modern datasets. All three relationships are tortuosity-cementation "m" variable in the same sense. All three are in relative "m" agreement considering the inter-granular tortuosity expected from their different insulating mineral grain aspect ratios.

Chapter-5.5 Porosity from Acoustic transit-time Sensor Input

Sonic logs are frequently used to estimate porosity from transit-time measurements. There is generally agreement between the velocity $\leftrightarrow$ porosity measurements made on laboratory cores at ultra-sonic frequencies and that observed with downhole logging sondes. However, almost every study shows a large mineralogic effect, particularly with respect to clay minerals, on measured acoustic velocities.

The propagation of compressional and shear waves through un-fractured sedimentary rocks closely follows the Extended Elastic Equations . The Gassmann (1951) equations, Woods equation and Hashin-Schtrikman (1963) equations, and Archer's (1992) NaCl brine relationships are all forms of Hooke's law (Holbrook, Goldberg & Gurevich, 1999b).

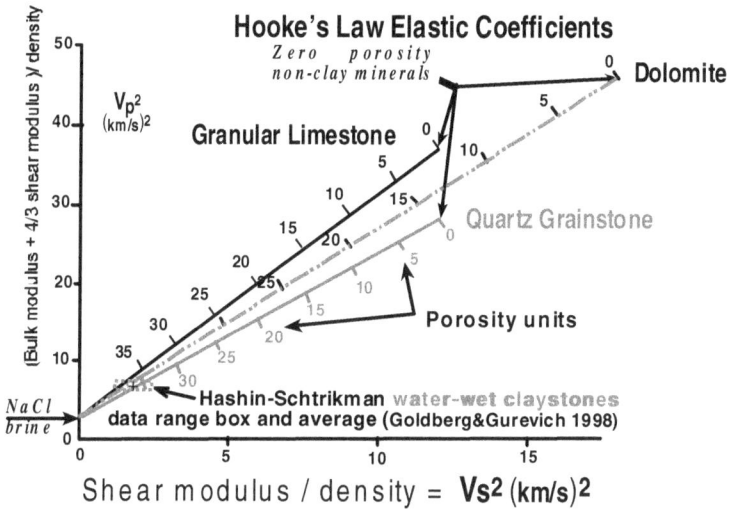

Extended Elastic Equation - Velocities (V_p^2 - V_s^2) crossplot of single mineral grainstones & claystones with $NaCl\ brine$

Hooke's Law Elastic Coefficients

Zero porosity non-clay minerals

Dolomite

V_p^2 (km/s)^2

Granular Limestone

Quartz Grainstone

Porosity units

(Bulk modulus + 4/3 shear modulus)/ density

$NaCl$ $brine$

Hashin-Schtrikman water-wet claystones **data range box and average** (Goldberg&Gurevich 1998)

Shear modulus / density = V_s^2 (km/s)^2

Wood's Equation	Gassmann Equations pore compliance($\alpha = \beta$)	Hooke's Law Coefficients
Initial 1.0 Grain Contact	0.0	are only Mineral Dependent

figure-5.2 shows the composite elastic coefficients – mineralogy – porosity – V_p^2 – V_s^2 relationships for the common sedimentary rocks. The V_p^2 and V_s^2 axes of the plot correspond to the Bulk Modulus (K), Shear modulus (μ) and Bulk density (ρ_{bulk}) terms in accordance with Gassmann (1951) granular solid elastic equations.

The measurement axes on figure-5.2 are V_p^2 and V_s^2. Adjacent to each axis is the Hooke's law coefficients, bulk modulus, shear modulus and bulk density that are equivalent to those squared velocities. The point being emphasized is elastic wave velocities for porous granular sedimentary rocks and slurries closely follow Hooke's law Extended Elastic Equations over that the entire (0% to 100%) porosity range.

The log datasets from which the claystone elastic coefficients were extracted had porosities ranging from 38% to 2%. All these lithologies were reasonably hard rocks, not slurries. The general slurry-like acoustic behavior for low porosity claystones is reasonable considering the acoustic wave travel-path on the molecular scale. Each clay particle has an associated inter-lammelar electrostatically bound water layer.

An elastic wave propagating through water-wet clay in any direction must pass through the much slower water phase in the inter-lammelar pore space. Even in very hard claystones, the individual clay lamellae are generally not in direct solid-solid contact. Water-wet claystone elastic behavior is overall slurry-like as each clay lamella is encased in its own electrostatically bound water layer. The effective elastic modulus of in $situ$ claystone elastic coefficients and velocities have been poorly understood.

To fill this information gap Goldberg & Gurevich (1998) performed a series of Hashin-Schtrikman inversions on mixed mineralogy V_p^2 and V_s^2 full waveform log datasets. Their average water wet end-member claystone elastic coefficients (velocities) falls into a very narrow range shown on figure-5.2. This is the same V_p^2 - V_s^2 region where elactrostatically neutral grainstones pass from a slurry suspension into an amalgamated granular solid.

Though a different physics is involved, there is a correspondence between the convergent formation factor region (37±3% porosity) on figure-5.2 with that Wood's equation to Gassmann equation transition region on figure-5.2. On both figure-5.1 and figure-5.2 the single mineral curves converge at the point where sediments of any mineralogy consolidate from a separate particle slurry to a granular solid.

The Extended Elastic Equations portrayed on figure-5.2 can be used to invert porosity from V_p^2 and/or V_s^2 in $situ$ petrophysical data when used in concert with petrophysically measured bulk density (ρ_{bulk}) and mineralogy (V_{clay}) from natural gamma-ray petrophysical data (Holbrook, Goldberg & Gurevich, 1999b).

Chapter-5.6 Conclusions on Petrophysical Sensor to Porosity Transforms

The γ–γ density, resistivity and acoustic transit-time sensors involve different governing physics. The physical properties of non-clay minerals are essentially constant under sub-surface geologic conditions.

The physical properties of NaCl brine vary with salinity and subsurface PVT according to Archer's (1992) Equation of State. The grain density of sedimentary clay minerals varies systematically with increasing temperature and pressure during diagenesis. These two variables can be systematically taken into account to improve porosities calculated from all petrophysical sensors.

Accurate porosity is probably the most important factor in reservoir evaluation and pore pressure determination. Porosity is both the mineral – fluid properties partitioning coefficient and a measure of absolute *in situ* strain (1.0 - φ). The earth's petrophysical – mechanical systems depends critically on this rock property. The estimation of oil in place, rock density, elastic moduli and rock strength are also critically dependent on rock mineralogy and porosity.

References cited

Aja, S.U., & P.E. Rosenberg, 1992, "The Thermodynamic Status of Compositionally-Variable Clay Minerals: A Discussion," Clays & Clay Minerals, vol. 40, p292.

Archer, D.G., 1992, "Thermodynamic properties of NaCl + H2O System II. Thermodynamic properties of NaCl(aq), NaCl.2H2O(cr), and phase equilibria," by J. Phys. Chem. Ref. Data, Vol. 21, No. 4, pp. 793-829.

Archie, G.E., 1941, "The Electrical Resistivity Log as an Aid in Determining Some Reservoir Characteristics", *Trans*; AIME, 14, 54

Borai, A. M., 1987, "A New Correlation of Cementation Factor in Low-Porosity Carbonates", SPE 14401

Carmichael, R.S., 1982, "Handbook of Physical Properties of Rocks", CRC Press

Gassmann, F., 1951, Ueber die Elistizitat poroser Medien: Viertelajahrschrift der Naturforschenden Gesellschaft in Zurich, v.96, p.1

Goldberg, I. & B. Gurevich, 1998, "Porosity Estimation from P and S sonic log data using a semi-empirical velocity-porosity-clay model", SPWLA 39th Annual Logging Symposium, paper QQ.

Hashin, Z., & S. Shtrikman, 1963, "A variational approach to the theory of the elastic behavior of multiphase materials", J. Mech. Phys. Solids, vol. 11, pp. 127-140.

Holbrook, P.W. I.Goldberg & B. Gurevich, 1999b, "Velocity - Porosity - Mineralogy Gassmann coefficient mixing relationships for water saturated sedimentary rocks". , paper T in SPWLA 40th Annual Logging Symposium, May –30 – June 3, 1999.

Huang, W.L., J.M. Longo, & D.R. Pevear, 1991, "An Experimentally Derived Kinetic Model for Smectite to Illite Conversion and Its Use as Geothermometer", presented at the 1991 Clay Minerals Society Annual Meeting, Houston Oct 5-10.

Hueckel, T. A., 1992, "Water-mineral interaction in hygromechanics of clays exposed to environmental loads: a mixture-theory approach", Canadian Geotechnical Journal V. 29, pp. 1071-1086.

Krief, M., Garat, J., Stellingwerf, J., and Ventre, J., 1990, "A petrophysical interpretation using the velocities of P and S waves (full waveform sonic), *The Log Analyst*, vol. 31, pp. 355-369.

Mao, Z.Q., C.G. Zhang, C.Z. Lin, J. Ouyang, Q. Wang, & C.J. Yan, 1995, "The Effects of Pore Structure and Electrical Properties of Core Samples from Various Sandstone Reservoirs in Tarim Basin". SPWLA 36 Annual Logging Symposium, June 26-29, 1995

Chapter-6. The related Extended Elastic and Grain-Matrix-Compactional MECHANICAL SYSTEMS.

The Earth's mechanical systems are composed of minerals and fluid. The solidity (1.0-φ) of granular sediment is a definition of absolute *in situ* grain-matrix strain. The minerals and fluids of the earth have discrete elastic, and grain-matrix-compactional stress/strain coefficients.

The closed-form grain-matrix-compactional mechanical system applies from initial grain contact to total consolidation. The stress/strain and density coefficients of the five most common sedimentary minerals have been experimentally determined (see table 1.1). NaCl brine compressibility and density coefficients are defined for all geologic PV/T conditions through Archer's (1992) Equation of State. The five most common sedimentary minerals and NaCl brine compose over 90% of all sedimentary rocks.

The mineral & fluid, stress/strain & density, coefficient-mixing laws are linear in the grain-matrix-compactional mechanical system domains. Overburden, pore pressure and fracture propagation pressure are related to mineralogy and porosity in the grain-matrix-compactional mechanical system in ≈biaxial Normal Fault Regime basins. They are also related in flat lying strike-slip fault bounded basins.

The individual terms of the elastic and Grain-matrix-compactional mechanical systems are mathematically related according to Hooke's law. The petrophysically measurable mineral-fluid-porosity compositional join is common to many other earth mechanical systems. Through these mechanical systems, pore pressure is related to mineral and fluid properties using modern petrophysical-mineralogical porosity transforms.

Chapter-6.1 The closed-form Force⇔balanced Mineral & Fluid properties solution to Pore pressure.

The earth's end-member mechanical systems are analogous to other known mechanical systems. Both end-member mechanical systems are force balanced constitutive relationships. Both system use only mineral, fluid density and stress/strain coefficients within their separately closed forms. The loading-limb and extended elastic equations mechanical systems are related through their common dependence on mineral , fluid and porosity rock physical properties. Newton's law and Hooke's law mechanical systems are both formulated using these rock properties as shown on the following page.

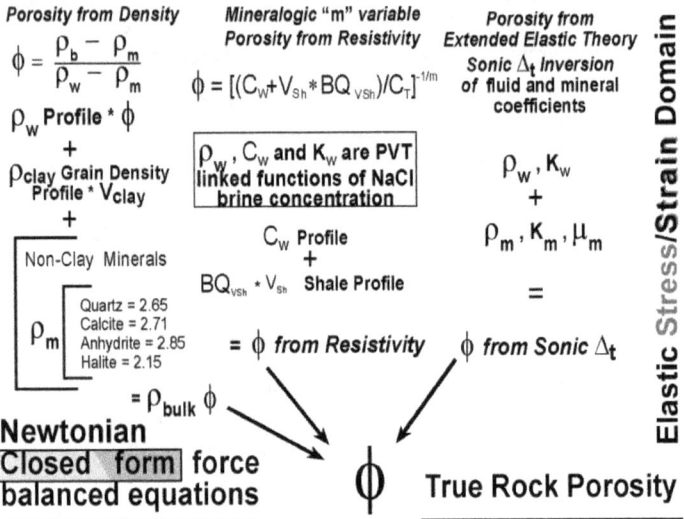

Figure-6.1. The extended _elastic_ and grain-matrix-compactional mechanical systems of the earth. The _elastic_ mechanical system is in the upper half of figure-6.1. **Porosity** and **mineralogy** are interpreted from the three petrophysical sensors shown. The elastic and density coefficients (K,μ, & ρ) vary with **porosity** and **mineralogy** in the Extended Elastic Equations mechanical system. The Sodium Chloride brine elastic and density coefficients (**K& ρ**) vary with NaCl salinity, pressure, and temperature according to the Archer (1992) Equation of State.

Chapter-6.2 The Extended Elastic Equations Mechanical Systems Domain

The upper half of figure-6.1 shows the elastic stress/strain domain for 100% water saturated sedimentary rocks. The rightmost column shows the average mineral and fluid coefficients that are entered into the Extended Elastic Equations mechanical system. Given an estimate of average mineralogy, porosity can be inverted from measured compressional or shear wave velocities through the Extended Elastic Equations mechanical system. Figure-5.2 shows this sedimentary rock V_p^2 vs. V_s^2 mechanical system.

True Rock Porosity is the connection between the elastic and grain-matrix-compactional mechanical systems domains. True Rock Porosity is the fluid-filled fraction of a porous sedimentary rock. The Extended Elastic Equations are the only velocity-porosity transform that represents a mechanical system. This system comprehensively accounts for clay and non-clay minerals and fluid constants that affect sedimentary rock V_p^2 and V_s^2 velocity. The Extended Elastic Equations cover the entire porosity range for all minerals from 0% to 100% porosity using a single mechanically representative system of equations.

True Rock Porosity is also calculated using a γ-γ bulk density sensor with knowledge of mineral and fluid density constants. The upper leftmost column on figure-6.1 shows the density equation and the source of the constants using subscripts and color codes. Bulk density (ρ_b) is measured. Mineral grain density (ρ_m) is the weighted average of the individual mineral bulk densities that compose the solid fraction of the rock. Water density (ρ_w) is calculated using a PVT–x NaCl transform derived from Archer's (1992) Equation of State.

The non-clay minerals have essentially constant bulk densities. The bulk density of clay minerals (ρ_{clay}) increases with increasing temperature and pressure. Water density (ρ_w) increases with salinity and pressure. In a given region water and average clay mineral bulk densities vary systematically with depth.

Equation-(5.1) is used to calculate the average grain density for a given mixed mineralogy rock. Archer's (1992) equation of state is used to calculate the density of Sodium Chloride brine under *in situ* PVT conditions. These two parameters are input as depth profiles to the bulk density-porosity transform (Equation-5.2) to achieve accurate True Rock Porosity results.

This bulk density-porosity transform accounts for all the mineral and NaCl brine density variability that is known to occur. Porosity errors up to 5 PU can occur if the wrong mineral grain density or water bulk density are used in the bulk density-porosity transform.

The central column in the upper half of figure-6.1 shows the Archie variable "m" methodology of calculating porosity from input resistivity sensor data. All resistivity-porosity transforms depend on the tortuosity of the electrical path through a porous sedimentary rock. Spherical grains have the lowest tortuosity. Clay minerals are high aspect ratio platelets. Claystones present a high tortuosity flowpath to Na^+ and Cl^- ions in their pore space.

Average mineralogy is used to interpolate between the end-member mineral formation factor vs. porosity transforms. The average rock formation factor will be between the Borai and claystone transforms in limestone-claystone lithostratigraphic sequences. The average rock formation factor will be between the Mao and claystone transforms in quartz grainstone-claystone lithostratigraphic sequences.

Formation water conductivity (C_w) varies with temperature and salinity. Geothermal gradients vary regionally. The salinity of connate formation waters in claystones usually varies gradually with depth and geologic formations. Formation water conductivity (C_w) that is input into the "m" variable resistivity-porosity transform is taken from a water conductivity vs. depth profile that is constructed for a particular region. Using this from the water conductivity profile, Equation-5.3 is used to calculate the Formation resistivity factor.

The three mineralogic end-member transforms are presently known to be curves on the power-law formation factor vs. porosity plane. Using these curves, porosity calculated from resistivity data input will match porosity calculated from the bulk density-porosity transform in the absence of borehole problems. Equation-5.4 expresses this non-linear Archie porosity – Formation factor transform.

Figure-5.1 shows the large non-linear mineralogic variability in the tortuosity factor "m". When "m" is properly accounted for equation-5.4 is used to calculate porosity from the Formation resistivity factor.

Archer's equation of state links NaCl brine salinity, bulk density (ρ_w), bulk modulus (K_w), and fluid velocity (V_w). The water physical properties profiles that are input into each of the three sensor-porosity transforms are physically consistent. Wherever (C_w), (ρ_w), or (K_w) are shown on figure-6.1, they are the same and are mathematically related to each other through the EOS governing physics.

A sedimentary rock has only one True Rock Porosity. This is estimated through a petrophysical sensor-porosity transform and transferred from the elastic to the grain-matrix-compactional mechanical system domain. If multiple sensors are available, porosities calculated from each sensor are compared. If the sensor specific porosity transforms match, one has increased confidence that the calculated porosity is the True Rock Porosity.

If the sensor specific porosities differ, there is something wrong in the sensor data input or the mineralogic constants that are input. Any discrepancy should be investigated and resolved. The three sensors shown depend on very different physics in their sensor-porosity transforms. The physically consistent overlaps provide a rational means to resolve any conflicts apparent in the data from different sensors.

Both mechanical systems depend on the same earth constituents. The elastic and compactional mechanical systems share the same True Rock Porosity, and bulk density constants. The same average mineralogy is also shared by the end-member mechanical system. There are many mineral and fluid constant overlaps shown in the upper and lower halves of figure-6.1.

Chapter-6.3 The force↔balanced grain-matrix-compactional system of equations

Force balance in the earth is vectorial within volumetric boundary conditions. In this physically representative solution, all the static solid vectors and the fluid scalars must individually and collectively sum to zero. The mineral and fluid density and stress/strain coefficients are in units that match stress, strain and pressure scalars and tensors. The mechanical system is dimensionally correct and physically representative.

True Rock Porosity and **mineralogy** from the elastic mechanical system domain feeds into the compactional **stress/strain** equation system in the lower half of figure-6.1. The compactional equation system is a mathematical extension of Newtonian physics within the effective stress theorem boundary conditions. Average effective stress is mathematically related to average *in situ* strain (**1.0-φ**) in the earth compactional mechanical system.

The diagonal background shading separates the internally force↔balanced confining **load** terms from the load-bearing (**grain matrix strain**) terms. All the earth **loads** are generated and borne by the earth's **minerals** and **fluid** that constitute sedimentary rocks. The equal (=) signs along the shading diagonal indicate the individual equation **load/strain** force balances. The equal (=) signs within a load or load-bearing domain are traditionally accepted mathematical definitions.

All the load bearing terms are related through their common **mineral** and **fluid** constant relationships to **volumetric grain matrix strain**. The Newtonian gravitational load generating term that is also Equation-1.1 has a simple positive relationship to **solidity** as well as **mineral** and **fluid** bulk densities. Earth **loads, volumetric strain,** and **solidity** are mathematically and physically related to **minerals** and **fluid** that generate and bear these loads through this closed-form equation system (Holbrook, 1998a, &1999).

Examination of the subscripted terms and algebra shown on the lower half of figure-6.1 reveals a highly redundant mathematical linkage of **loads, vertical,** and **horizontal effective stresses.** Pore pressure is mathematically related to **strain (1.0-φ)** and **mineralogy** within the compactional closed-form system of equations.

This closed-form equation system is the simplest possible combination of mineralogic **effective stress/strain** and density coefficients, with **load,** and **strain** definitions. This grain-matrix-compactional mechanical system also uses and explains **fracture propagation pressure** as an accompanying minimum-load **stress/strain** relationship.

The dark gray vertical arrow through the load-bearing terms emphasizes the direct linkage of **solidity** to the partial load terms **pore pressure** and **fracture pressure. Solidity** is a gravitational load generating term that is mathematically equivalent to **grain-matrix strain** by definition. This (**solidity**=**strain**) definition equivalence contributes to the mathematical simplicity of this earth closed-form mechanical system.

The definition of **grain-matrix strain** equals (**1.0-φ**) is the first equation in the closed-form grain-matrix-compactional mechanical system. That equation is entirely within the darker gray strain region of figure-6.1. figure-1.1 shows this definition as a partitioning of a decreasing fluid volume with a constant solid volume.

On figure 1.1, an arbitrary volume of fluid and solid are initially deposited on the seafloor. In this example that arbitrary solid volume is ½. In three ½ steps, the volume of fluid in that initial sediment is expelled due to compaction. Solidity increases from ½ to ¾ to 7/8 in these three ½ steps. The grain framework compactional strain limit is reached when all of the fluid is expelled from the initial sediment or sedimentary rock. At this limit, solidity equals 1.0.

Solidity is a valid definition of **strain** if no solid matter is added during compaction. Even if some solid material is added or subtracted from a sedimentary rock, its altered mineralogy is the effective stress load-bearing element. Providing that the altered mineralogy is known, loading-limb effective stress can still be estimated from mineralogy and strain.

Newtonian gravitation **load** (S_v) is expressed in terms of load-generating **mineral** and **fluid** density (ρ) coefficients in the second equation on figure-6.1 that is also equation-1.1. The density coefficients are partitioned by **solidity**. The equal (=) sign in the Newtonian gravitational equation is along the diagonal that mathematically links **load** (S_v) to *in situ* **solidity** and **strain**. Equation-1.1 in chapter-1.0 gives a more thorough explanation.

The third mechanically representative equation is the First Fundamental in situ Stress/Strain Relationship. Figure-1.2 and equation-1.2 describe this power-law compactional relationship for the five most common sedimentary minerals. Average effective stress is mathematically related to volumetric *in situ* strain through two mineralogic constants (α & σ_{max}). Alpha is the power-law slope of the **stress/strain** relationship. Sigma max is the average effective stress at the grain-framework-compactional **strain** limit (solidity=1.0). Holbrook (1995) demonstrated that normal single mineral compaction curves are represented by force balanced, power-law linear **stress/strain** mathematical functions.

On figure-1.2 the hardest common sedimentary mineral, quartz, has the highest compaction resistance (σ_{max}), and the softest, halite, has the lowest. The power-law exponent (α) for end-member claystones incorporates inter-particle repulsion. The non-clay minerals have sub-parallel compaction exponents. Non-clay minerals resist compaction only through their crystalline lattice. Clay mineral particles have a compactional resistance above that of their hardness due to their inter-particle repulsive force (See chapter-2.0).

The fourth equation on figure-6.1 is the Second Fundamental in situ Stress/Strain Relationship (equation-4.2). As the vertical effective stress load (σ_v) increases, the horizontal effective stress load (σ_h) also increases in direct proportion to volumetric strain. Figure-4.1b shows the *in situ* strain relationship in comparison with some empirical (σ/σ_v) depth functions (figure-4.1a.) reported by Pilkington (1978).

The empirical stress ratio (σ_h/σ_v) vs. depth relationships (figure-4.1a.) and the physically representative *in situ* strain vs. depth function (figure-4.1b) have similar shapes and magnitudes.

The **solidity** data points on (figure-4.1b) were fit with the single power-law relationship that is shown. The fit is excellent. There is extensive discussion of these figures in Chapter-4.0.

Fracture pressures and leakoff tests predicted using the Second Fundamental stress ratio (σ_h/σ_v) = *in situ* strain relationship are about four times more accurate than fracture pressures predicted from empirical depth functions (Holbrook, 1989, 1996). This simple mechanically representative theory is supported by much more accurate leakoff test prediction results.

The fourth equation is a definition of biaxial in a Normal Fault Regime basin. Many of the horizontal stress anisotropies that are observed in these basins are related to local fault boundary conditions and structural dip (Holbrook, 1999). These phenomena are usually of a low magnitude. Any tendency toward the development of increased stress anisotropy is buffered toward the biaxial 2:-1:-1 even number relationship (equation-4.2 & figure-4.2). Pressure solution would be greater in the direction of greater horizontal stress tending to minimize horizontal stress anisotropy (chapter-4.4).

The fifth equation in the closed formulation is the effective stress theorem ($S_{ave} - \sigma_{ave} = P_p$). This equation is a global scalar force↔balance definition. The scalar **pore pressure** is calculated exactly through the subtraction of two equivalent scalars ($S_{ave}-\sigma_{ave}$). This properly matches units and honors the effective stress theorem, equation-1.2 boundary condition.

Fracture propagation pressure ($\mathbf{P_F = P_p + \sigma_h = S_h}$) is a load definition calculated in the sixth equation. It uses **pore pressure** ($\mathbf{P_p}$) calculated from the effective stress theorem definition. Effective stresses are force balanced with respect to each other by definition. All the light gray background load terms on figure-6.1 are a Newtonian closed-form force balance!

All the earth load-bearing strain terms have a light gray background. Grain-matrix volumetric *in situ* strain ($1.0 - \phi$) is in each of the individually force balanced **stress/strain** equations. Loads from any direction contribute to volumetric **strain**. The descending arrow on figure-6.1 indicates the algebraic linkage of each equation to measurable *in situ* (**solidity=strain**).

Sigma max and alpha (σ_{max} & α) are volumetric **stress/strain** coefficients of the effective stress load-bearing **minerals**. Force↔balanced effective stress vs. grain-matrix **strain** linkage is unique to this closed-form Newtonian formulation and leads to simplicity and accuracy of calibration, prediction and detection of **pore pressure**.

Figure-4.2 summarizes the force↔balanced relationship between the First and Second *in situ* Stress/Strain relationships. The compactional average effective stress/volumetric *in situ* strain relationship is shown as a solid power-law linear effective stress/strain relationship. The mineralogic sigma max and alpha (σ_{max} & α) are that of an end-member claystone. The stress ratio (σ_h/σ_v) and *in situ* strain relationship with respect to average effective stress are shown as short dash and long dash power-law linear relationships. Average, minimum, and maximum effective stresses are biaxially force↔balanced.

These relationships honor the effective stress theorem and explain the empirical Terzaghi (1923) Effective Stress "Law" for marine sediments. Terzaghi's "law" is a mixed uniaxial vector–scalar relationship with the horizontal vectors undefined. Gravitational (σ_v) and biaxial horizontal (σ_h) effective stresses vary in direct proportion to each other in the closed-form earth mechanical system. This is the only mechanically sensible explanation as to how Terzaghi's "law" has been successful for so many years at predicting pore pressure (Holbrook, 1997).

Fracture propagation pressure ($P_F = P_p + \sigma_h = S_h$) is also mechanically explained by these relationships. For a single mineralogy, like the end-member claystone example on figure-4.2. Fracture propagation pressure is a single force$\leftrightarrow$balanced power-law linear effective stress/strain relationship that is described by equation-4.2.

Leakoff tests in limestones are generally higher and leakoff tests in quartz grainstones are generally lower. These systematic mineralogic differences are mechanically explained by the accompanying offset of mineralogic coefficients (σ_{max} & α). The lower (**solidity=strain**) of quartz grainstones decreases fracture propagation pressure. The higher (**solidity=strain**) of limestones increases fracture propagation pressure.

Fracture propagation pressure (S_h) is affected by both variable minimum horizontal stress and pore pressure ($P_F = P_p + \sigma_h$). The closed-form compactional mechanical system explains both the single mineral fracture trends and the mineralogically variable fracture pressure differences as changes in (**solidity=strain**). A mechanically sensible theory that explain these similarities and differences is that geologic loading rate compaction for all sedimentary minerals is essentially plastic (Holbrook, 1996).

The mechanical behavior of natural sediments at their upper (σ_{max}) and lower limits are plastic. The separate loading-limb stress/strain functions between these limits on figure-4.2 are all power-law linear. This power-law linearity between the upper and lower limits is consistent with equilibrium compaction of a grain-matrix-framework. That compaction being primarily dependant on the average mineralogy of the sediment being loaded (Holbrook, 1995).

Chapter-6.4 The *in situ* stress/stress relationship for strike-slip basins

The third, fourth, and fifth equations on figure-6.1 are appropriate for $\cong$biaxial Normal Fault Regime tectonic regimes. The relationships between principal and average effective stresses are different in strike-slip tectonic regimes.

Equations-4.3 [$\sigma_{ave} = \sigma_v$, $\sigma_h = \frac{1}{2}\sigma_v$, and $\sigma_H = 3/2 \sigma_v$] should replace The fourth, and fifth equations on figure-6.1 for strike-slip fault regime tectonic settings. Overburden is calculated using Newton's law that is the second equation of figure-6.1 in both mechanical systems. Pore fluid pressure and fracture propagation pressure (the sixth and seventh equations on figure-6.1) are the same in both ≅biaxial Normal Fault Regime and Strike-slip fault regime mechanical systems.

This resultant closed-form mechanical system depends on the overburden – vertical stress equivalence. The porosity – strain correspondence evident in ≅biaxial NFR mechanical systems was not exactly equal in the strike-slip mechanical system (see figure-4.5). However, equation-4.2 contain the average effective stress relationship through the effective stress theorem. The strike-slip fault regime mechanical system is effectively a closed-form because of the mechanical fracture pressure/pore pressure limits described in chapter 7.

Stike-slip tectonic regimes have sheared sub-vertical fracture zones. The minimum confining stress is usually near or below hydrostatic fluid pressure. Because of the minumum principal stress is related to the vertical according to ($\sigma_h = \frac{1}{2} \sigma_v$) and the vertical confining load is overburden, pore pressure is usually near hydrostatic in strike-slip fault regimes.

Chapter-6.5 Geologic, geotechnical, and engineering applications of earth mechanical systems.

Earthvborehole force balance is a primary concern in most sub-surface engineering operations. The earth load state terms, effective stresses, rock elastic and density properties, are dependent on **porosity,** and **mineralogy.** The inter-dependencies of the Extended Elastic and *in situ* grain-matrix-compactional mechanical systems offer significant engineering advantages. The physically representative, quantitative mechanical systems outputs are listed by type on table-6.1 through table-6.5 below.

The mechanical systems inter-dependence allows simultaneous calibration of 1.) borehole fluid pressure s; 2.) relative mud weight, 3.) formation fluid pressure tests, and 4.) leakoff tests to the corresponding physically representative earth mechanical system (Holbrook, 1996). These are all corresponding unit calibrations between borehole measurements and the two closed-form earth mechanical systems.

Pore pressure , and fracture pressure between measurement points and in nearby wells can then be predicted using the physically calibrated closed-form compactional mechanical system. These same mechanical system variables and rock elastic properties are also important to subsequent borehole mechanical operations. Setting casing, well completion, and reservoir engineering all depend on these same loads and physical rock properties.

Rock properties from the Extended Elastic Equations mechanical system and rock load state data from grain-matrix-compactional mechanical systems can be applied simply and directly to solve drilling, reservoir, and completion subsurface engineering problems.

Quantitative Rock Properties and Loads output for each foot of petrophysical data

Table-6.1. Constitutive rock or sediment properties

1.) **Porosity; Solid volume fractions** 2.) Clay minerals 3.) Quartz 4.) Calcite 5.) Halite

Table-6.2. Whole rock elastic and density properties

6.) Bulk modulus 7.) Shear modulus 8.) Young's modulus 9.) Bulk density 10.) Compressional wave velocity 11.) Shear wave velocity 12). Dry rock Poisson's ratio

Table-6.3. NaCl brine properties

13.) Electrical conductivity 14.) Density 15.) Compressional wave velocity

Table-6.4. Rock confining Load State and Effective stress data

16.) Pore fluid pressure 17.) Overburden= Maximum vertical load 18.) Fracture propagation pressure = minimum horizontal load 19.) Average effective stress 20.) Vertical = maximum effective stress 21.) Horizontal minimum effective stress

Table-6.5. Regional temperature related profile data

22.) Geothermal gradient 23.) NaCl brine conductivity 24.) Clay mineral grain density

References cited

Archer, D.G., 1992, "Thermodynamic properties of NaCl + H2O System II. Thermodynamic properties of NaCl(aq), NaCl.2H2O(cr), and phase equilibria," by J. Phys. Chem. Ref. Data, Vol. 21, No. 4, pp. 793-829.

Holbrook, P W, 1995, "The relationship between Porosity, Mineralogy and Effective Stress in Granular Sedimentary Rocks", paper AA in **SPWLA 36th Annual Logging Symposium**, June 26-29, 1995.

Holbrook, P W, 1996, "The Use of Petrophysical Data for Well Planning, Drilling Safety and Efficiency ", paper X in **SPWLA 37th Annual Logging Symposium**, June 16-19, 1996.

Holbrook, P W, 1997, "Discussion of A New Simple Method to Estimate Fracture Pressure Gradients", **SPE Drilling & Completions**, March 1997, pp.71-72

Holbrook, P W, 1999, "A simple closed-form force balanced solution for Pore pressure, Overburden and the principal Effective stresses in the Earth.", **Journal of Marine and Petroleum Geology**, Vol. 16, pp. 303-319.

Terzaghi, K. Van, 1923, "Die Berchnung der Durchassigkeitziffer des Tones aus dem Verlauf der Hydrodynamischen Spannungscheinungen", Sitzunzsber Akad Wiss. Wein Math Naturwiss, K1.ABTS 2a, pp. 107-122.

Chapter-7. Fracture Pressure limit to *in situ* Pore Pressure Profiles.

There are complementary mechanical systems relationships in addition to the physically closed-form mechanical systems discussed in the first 6 chapters. Earth Force balance, Pascal's principle, and Darcy's law flow simultaneously regulate Pore pressure profiles. Darcy flow occurs along the least work path from any point in the subsurface to the Earth's surface. Fluid flow rate is regulated by the net (fracture + intergranular) permeability of the Earth formations. At natural geologic flow rates, any continuous lithostratigraphic unit with over 10 millidarcies net permeability is essentially at rest. The fluid pressure gradients within these essentially static fluid "pressure compartments" can be closely calculated from ($\rho \cdot g \cdot h$) Pascal's principle.

The fluid pressure at the minimum work leakpoint ($P_{p@lp}$) of any "pressure compartment" governs the fluid pressure of the entire compartment in accordance with Pascal's principle. That pressure is;

$$P_p = P_{p@lp} \pm [(\rho \cdot g \cdot h) \cdot \Delta h] \qquad (7.1)$$

The minimum work leakpoint of any pressure compartment could be an open fault which cuts the compartment or a closed fracture somewhere in the pressure compartment's overlying caprock. The caprocks overlying pressure compartments are characterized by very low intergranular permeability, very low porosity, and very high minimum principal stress.

In compacting sedimentary basins, the least work fluid escape flowpath from great depth to the surface can be characterized as a set of pressure compartment s with intervening stress sensitive pressure "seals". Minimum principal stress ($S_{min} = P_F$) governs the release of compartment fluid pressure regardless of pressure generating mechanism. The minimum work leakpoints are pressure gradient nodes on the Darcy law least work flowpath. The minimum work leakpoints determine the upper limit caprock "seal" relative hydrostatic of fluid pressure within "pressure compartments". Minimum principal stress at the minimum work leakpoints is the upper limit regulator of single fluid phase Darcy flow and pore pressure everywhere in the subsurface.

Chapter-7.1 Force Balance regulation of Compartment *in situ* Pore Fluid Pressure by Sealing Caprock *in situ* Fracture Pressure

Everywhere in the subsurface Darcy flow is limited by net (intergranular + fracture) permeability. Intergranular permeability is governed by grain matrix properties while fracture permeability is controlled by *in situ* stress. The minimum principal stress at the minimum work leakpoint determines the upper limit of fluid pressure within a "pressure compartment ".

Chapter-7.1.1 Fracture pressure dominates net permeability

Fracture permeability is the overwhelmingly dominant component of net permeability as the effective stress normal to minimally stressed fracture planes approaches zero (Walsch, 1981). The net permeability of a caprock pressure "seal" is highly stress sensitive. For most rocks, fracture permeability is conservatively 100X greater than the rock matrix intergranular permeability (Lorenz, et al, 1989). At this ratio or higher, intergranular permeability is a negligible (<1%) portion of net permeability.

Figure-7.1, shown on the facing page was taken from (Walsch 1981). It shows the extreme sensitivity of fracture permeability to minimum confining stress. Walsch's empirical equation is supported by several types of data from well fluid injection pressure and well drawdown pressure procedures shown on figure-7.1. His semi-empirical equation relates relative permeability (K/ K_0) to effective stress as shown on the figure. The horizontal axis of figure-7.1 is effective stress normal to minimally stressed fissures in the rock. Minimum effective stress varies from zero to 6000 psi on the horizontal axis.

The logarithm of net permeability is shown on the vertical axis as a ratio. The Walsch equation shown on figure-7.1 is normalized to the net permeability (K_0) at the initial Reservoir State before any fluid is injected or withdrawn. The highest *in situ* measured net permeabilities are from nitrogen injection fracturing procedures.

Fracture permeability increases as a steep power-law function as fluid pressure approaches formation's fracture propagation pressure (Walsh, 1981, figure 1). Consequently, the net permeability which controls Darcy flow of fluids to the mudline is dominantly the fracture permeability as it is (100X to 10,000X) greater than intergranular permeability. The over-riding stress dependence of fracture permeability makes fracture propagation pressure the dominant regulator of compartmentalized pore fluid pressure .

Force balance ⊕ regulation of Net permeability
and pore fluid pressure profiles

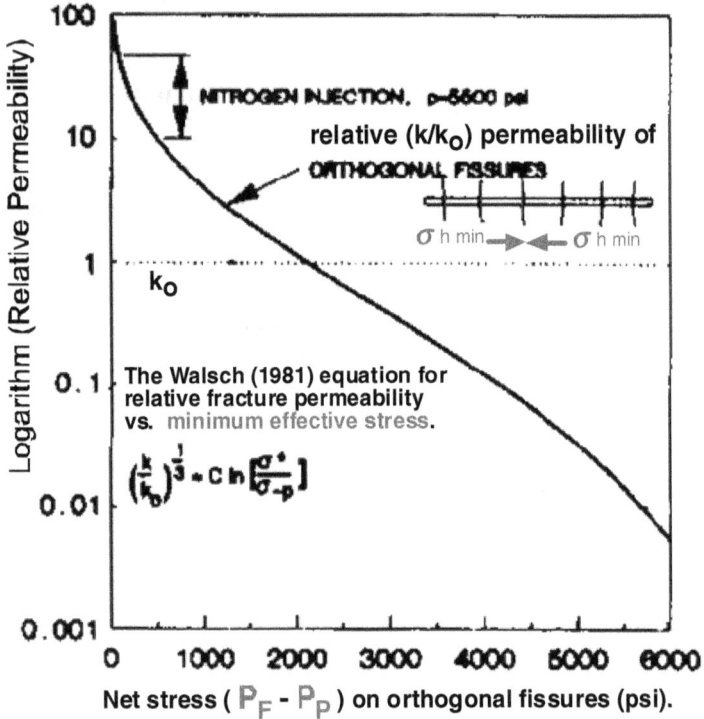

Figure-7.1 Net (intergranular + fracture) permeability to Darcy flow increases exponentially as pore fluid pressure approaches fracture propagation pressure. Adapted from Walsch (1981). fracture propagation pressure is the upper limit in situ pore fluid pressure regulating mechanism for single-phase fluids on a Geologic time scale.

Chapter-7.2 Relative fracture permeability is independent of fracture opening mechanism.

The permeability that limits the pore fluid pressure within a relative hydrostatic pressure compartment is the caprock fracture permeability. Engelder & Lacazette (1990) presented a poroelastic model indicating that fractures within a permeable caprock would be opened by elastic grain contraction when the pore fluid pressure exceeds the fracture propagation pressure of the caprock.

If there is zero permeability and no fluid penetration into the caprock the mathematical force balanced conclusion is the same. In either case, minimum principal stress holds the least work fracture closed and operates as a natural *in situ* pressure relief valve. The difference between the permeable vs. impermeable formation pressure dependent relationships is where the fluid pressure force balance is applied.

Fracture pressure is the force balance regulator of pore pressure irrespective of mechanism. Fractures normal to the minimum principal stress will be opened when pore pressure in the fracture slightly exceeds fracture propagation pressure. In compacting sedimentary basins, Darcy's law dictates that if pore fluid pressure is greater below the caprock than above; fluid will flow through the open fracture in direct response to that fluid pressure gradient.

Chapter-7.3 General "seal" fracture pressure / "compartment" pore pressure relationships

Figure-7.2 shows a generic pressure compartment illustrating the compartment - sealing caprock force balance inter-relationships which control *in situ* pore fluid pressure. The stippled area between the two fractured shale beds represents a continuous seal relative hydrostatic fluid pressure compartment. This compartment could be any shape or dimension. Its fluid pressure profile will be leakpoint relative hydrostatic obeying Pascal's principle. Fracture pressure at the pressure compartment free water level is the upper limit for the seal relative hydrostatic fluid pressure gradient within a continuous pressure compartment .

Generic continuous pore fluid Pressure Compartment showing the caprock fracture ⊕ pressure compartment pore ↓ pressure limit system.

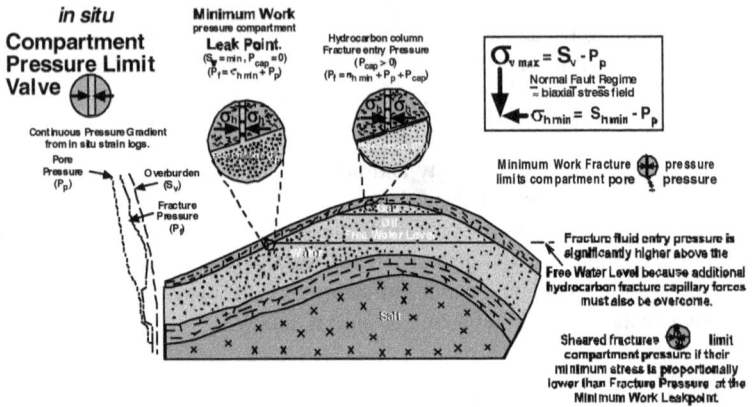

in situ **Compartment Pressure Limit Valve** ⊕↓

Continuous Pressure Gradient from in situ strain logs.

Pore Pressure (P_p)

Overburden (S_v)

Fracture Pressure (P_f)

Minimum Work pressure compartment **Leak Point.** $(S_v = min, P_{cap} = 0)$ $(P_f = \sigma_{h\,min} + P_p)$

Hydrocarbon column Fracture entry Pressure $(P_{cap} > 0)$ $(P_f = \sigma_{h\,min} + P_p + P_{cap})$

Free Water Level

Salt

$\sigma_{v\,max} = S_v - P_p$
Normal Fault Regime ≈ biaxial stress field
$\sigma_{h\,min} = S_{h\,min} - P_p$

Minimum Work Fracture ⊕ pressure limits compartment pore ↓ pressure

Fracture fluid entry pressure is significantly higher above the Free Water Level because additional hydrocarbon fracture capillary forces must also be overcome.

Sheared fractures ⊕↓ limit compartment pressure if their minimum stress is proportionally lower than Fracture Pressure at the Minimum Work Leakpoint.

Figure-7.2 Generic continuous pore fluid pressure compartment showing the caprock mated fracture pressure valve system. Caprock minimum work fracture pressure is the upper limit for seal relative fluid pressure within the compartment. Representative force balanced in situ Rock Mechanics System continuous logs are shown to the left.

The mated fracture minimum work leakpoint is at the free water level because additional hydrocarbon capillary forces must be overcome for two phase fluid entry into pore size fractures.

Chapter-7.4 Minimum Work Leakpoint location with respect to force balance

Elevated pore pressure at the minimum work leakpoint will open fractures in the compartments sealing caprock. Compartment fluid will easily escape through these fractures as long as they are held open by compartment pore pressure. Compartment pore fluid pressure will be bled down rapidly until the caprock fractures close. This occurs when the pore pressure in the fractures falls below the caprock fracture propagation pressure.

This spatial minimum *in situ* fracture pressure / pore pressure force balance limiting relationship applies under all tectonic conditions and is thus general in the subsurface. Borehole leakoff tests measure this pressure if they were conducted over a sufficiently long time.

An inset circle in the upper left of figure-7.2 represents a single vertical fracture perpendicular to the minimum principle stress within the caprock. Fracture pressure is portrayed as an *in situ* Compartment Pressure Limit Valve, which exactly describes its means of operation in the subsurface. The two opposing arrows in all 3 inset circles represent the minimum principal stress in ≈ biaxial Normal Fault Regime Basins. This force holds the faces of the pressure relief valve closed.

A mated sub-vertical fracture with no shear offset will be closed if the pore fluid pressure within the fracture is less than or equal to the caprock fracture propagation pressure. When the pore pressure in the compartment, which is exposed to the valve, exceeds the fracture propagation pressure, the valve will open. Fracture permeability increases exponentially as pore pressure approaches caprock fracture propagation pressure (Walsch, 1981). As fluid passes through the open valve, fluid pressure within the compartment will fall until the fracture closes at fracture propagation pressure.

Chapter-7.5 Additional capillary pressure for two phase fluids

The Minimum Work pressure compartment leak point shown on figure-7.2 is just below the hydrocarbon water contact. If there are no separate phase hydrocarbons in the pore fluid, and the caprock has uniform petrophysical properties, the Minimum Work Leak Point is at the highest elevation of the pressure compartment. If oil or gas is present as a separate phase in the pore fluid, the two phase fluid capillary entry pressure is always many times higher than the single phase capillary entry pressure. It takes much more work to force a two phase fluid into a capillary size fracture or pore than for a single-phase fluid. The center inset circle on figure-7.2 shows that the additional capillary entry pressure must be overcome for the two phase fluid to enter the capillary size fracture.

Chapter-7.6 Systematic Earth gravitational relative fracture pressure / pore pressure relationship.

The force balance at the pressure compartment - caprock interface changes systematically with overburden and elevation in figure-7.2 as it does with any pressure compartment . The fluid pressure within a pressure compartment changes in direct proportion to average fluid density / elevation (Pascal's principle). For subsurface brines this fluid pressure gradient is somewhere between 0.434 to 0.507 psi/foot. The change in Fracture pressure and overburden with elevation is somewhere within the range of 0.8 to 1.15 psi/foot. Fracture pressure gradient is about two times greater than relative hydrostatic pore pressure gradient along the upper caprock - compartment interface.

For each foot of elevation gained at the compartment - caprock interface; the relative (Pf - Pp) pressure holding a fracture closed decreases by about 0.4 to 0.6 psi. Other factors being equal the shallowest single-phase fluid in a compartment can open a mated subvertical fracture in a caprock with the least work. With increasing fluid pressure within a pressure compartment , fractures at this Minimum Work location would open first at that fracture pressure. These fractures would remain open so long as the pore pressure within the fracture is equal to or greater than the seal fracture pressure (Engelder & Lacazette,1990).

Thus force balance, (caprock fracture pressure - pore pressure) is the dominant pressure valve like regulator of pore fluid pressure in the subsurface. Like a pressure relief valve, fracture permeability is highly stress sensitive such that any additional increase in compartment pore pressure will be accompanied by a power law increase in fracture permeability. This *in situ* force balance pressure relief valve system will tend to be limited by the Minimum Work caprock fracture pressure leakpoint on a geologic time scale.

Chapter-7.7 The effect of active shear faults on minimum principal stress

If a pressure compartment is cut by a fault that has experienced recent shear displacement, the compartment fluid pressure upper limit would probably be lower. Stress is not evenly distributed along a fault plane immediately following a shear displacement. Relatively lower stress surface irregularities along the fault plane would probably have higher fracture permeability and thus offer less overall flow resistance to the mudline. If these higher permeability zones along the fault plane have some physical continuity, they would temporarily become the Minimum Work fluid flowpath.

However, with time the lower stressed voids will tend to be filled by plastic grain movement and chemical deposition of cement from waters passing through the fracture. The natural tendency is for partially open fractures to close and for the compartment pore pressure to gradually rise back up to the caprock Minimum Work fracture pressure.

Chapter-7.8 Basin general limit relationship, Northern North Sea example

Figure-7.3 illustrates the general *in situ* seal fracture pressure / compartment pore pressure limiting relationship. An extensive set of leakoff tests compiled by Gaarenstroom (1993) in the Central North Sea are shown as boxes. The crosses are Repeat Formation Test measurements made in this same area. The minimum fracture propagation pressure vs. depth trend was the hand drawn by Gaarenstroom.

The slope of the two dashed lines on the left represent the maximum and minimum expected hydrostatic gradients for water saturated rocks. The expanded inset box shows a classic example of a set of RFT measurements (+) made within a continuous pressure compartment . Their hydrostatic slope is within the expected hydrostatic range. The shallowest RFT is just below Gaarenstroom's minimum leakoff test pressure vs. depth trend line. The caprock seal at this depth can hold no more than the immediately underlying shallowest (minimum work) pore pressure in the pressure compartment .

Many other sets of reservoir RFT's can be recognized on the figure-7.3 compilation. All of them terminate below the Minimum Leakoff Pressure Trend line shown. These measurements agree entirely with the general fracture pressure/pore pressure force balance limiting relationships described above. This same fracture pressure/pore pressure force balance limiting relationship has been observed in many other basins using different datasets (Grauls, D., 1996). Nashaat (1998) has shown similar leakoff test fracture pressure/pore pressure relationships in the Nile Delta.

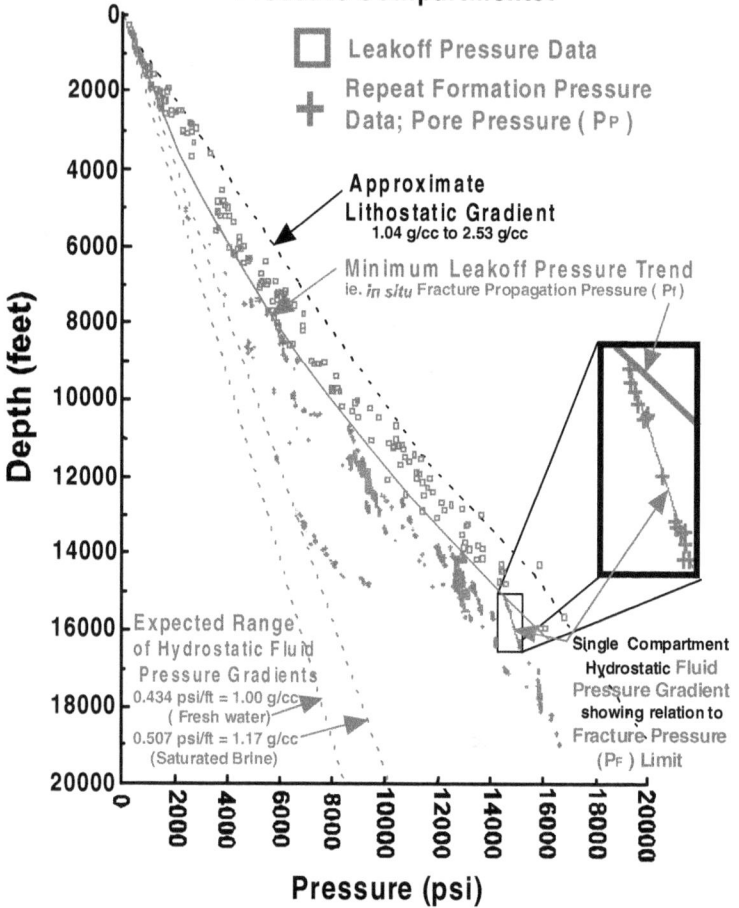

Fracture Pressure ⊕ is the Upper Limit of
Pore Pressure in Central North Sea
Pressure Compartments.

Figure-7.3 Leakoff test   and Repeat Formation Test + data from the
Central North Sea. The minimum leakoff pressure trend limits the pore
fluid pressure in all the pressure compartment s. The enlarged inset box
demonstrates the general **pore pressure ≤ fracture pressure** limiting
relationship. Fresh water and saturated NaCl brine fluid pressure
gradients are shown for perspective. Adapted from Gaarenstroom et al
(1992).

Chapter-7.9 Fracture pressure dominates all pressurization mechanisms

The thermal cracking of hydrocarbons is the most agressive fluid pressurization mechanism. Figure-7.4, taken from Barker (1990) is in the same pressure vs. depth format as figure-7.3. The phase change associated with the natural subsurface cracking of oil to gas dramatically increases peak potential pore fluid pressure. The cracking of oil to gas from 12,000 to 17,000 feet in figure-7.4 would raise pore fluid pressure to 5 times hydrostatic given a perfect seal.

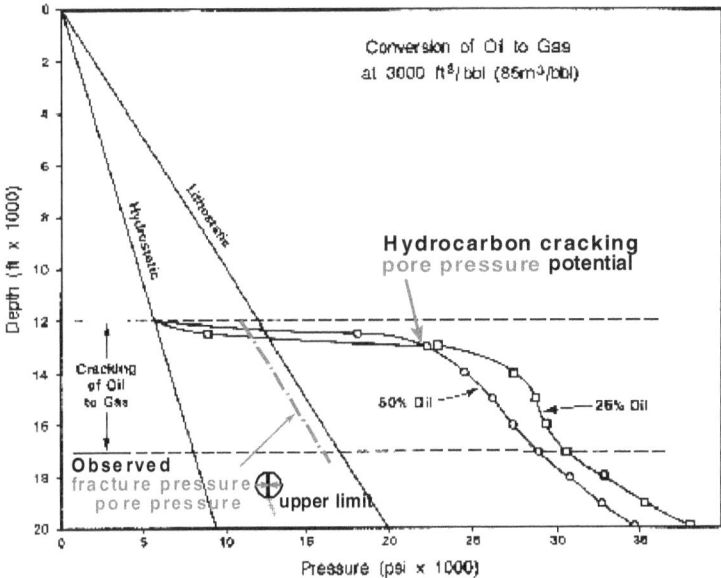

Figure-7.4. Pore pressure potential from the cracking of oil to gas between 12,000 and 17,000 feet. The PVT fluid volumetric expansion would generate pore fluid pressures that are five times hydrostatic and two times lithostatic given a perfect caprock seal.

Oil does crack to gas in this temperature - depth range. Fluid pressures above lithostatic are not observed in Normal Fault Regime ≈biaxial basins. The effective control of in situ pore pressure profiles is pressure seal capacity. Seal capacity limits even the highest fluid expansion un-loading limb excess pore pressures generating potential. Taken from Barker (1990).

Caprock seals are pressure sensitive as has been shown above. Thermal cracking fluid expansion is a significant pressurization mechanism in the North Sea data portrayed in figure-7.3. There are steep fluid pressure gradients associated with regional "seals" (Ward, C., 1995, 1996). The regional "seal" crosses stratigraphic boundaries, but is recognized through its low porosity. Low porosity is force balance proportional with high fracture pressure in Normal Fault Regime Basins (Holbrook, P.W., 1992,1996,1997,1998,1999).

The steep fluid pressure ramps begin at the regionally continuous low porosity - high fracture pressure "seal". But, in no place does the peak fluid pressure exceed fracture propagation pressure at minimum-work leakpoints on figure-7.3. The general conclusions one could reach through comparison of figure-7.3 and figure-7.4 are; Pressure seal capacity completely dominates the most aggressive fluid pressurization mechanism. Pore fluid pressure is limited by fracture propagation pressure in water saturated sedimentary rocks.

Chapter-7.10 General force balanced *in situ* Darcy flow relationships

Both natural brine and partially gas saturated ($\rho \cdot g \cdot h$) seal relative static fluid pressure profiles are highlighted with a pink pattern. Intervening steeper pore fluid pressure gradient indicating pressure drop due to Darcy flow are highlighted with a gray pattern. Pore fluid pressure profiles and Darcy flow are successively governed by one of the flow regulating mechanisms portrayed in the caption box depending upon which is least work. All the regulation mechanisms are minimum work relationships relating net permeability to minimum work Darcy flow. Mated fractures are held closed by minimum principal stress (S_{min}) along their whole length. The level of this stress is a function of overburden and solidity in Normal Fault Regime ≈biaxial basins (Holbrook P, 1996).

86

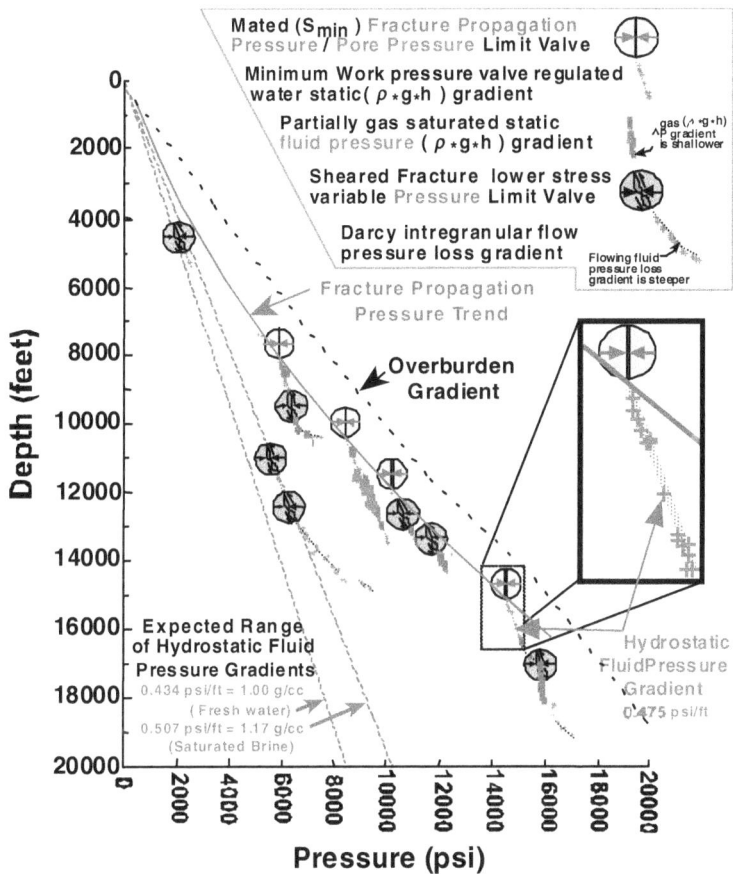

Figure-7.5 Dependence of pore fluid pressure gradient profiles on minimum work leakpoints and intergranular permeability. The relationships between mated fractures, sheared fractures, and three types of fluid pressure gradients are shown in the inset caption box. This figure-was adapted from Gaarenstroom et al (1992).

The stress distribution across recently sheared fractures is uneven. Local stress is highest at apical points of contact across the fracture. The minimum stress normal to a sheared fracture will often be less than the regional minimum principal stress (S_min) because of the uneven stress

87

distribution. Local stress drops to zero in sheared fracture zone voids. Given any degree of void hydraulic connectivity, sheared fractures will have higher permeability than mated fractures. These two types of stress-sensitive pore fluid pressure release valves are illustrated in the caption and the diagram of figure-7.5.

Pressure loss due to Darcy flow is seldom measured through RFT comparison in the subsurface. The few case examples where this is demonstrated are highlighted with a black pattern on the caption and figure. These pressure gradients are steeper than the water and partially gas saturated static ($\rho \cdot g \cdot h$) fluid pressure gradients that are normally observed through RFT comparison.

The term "pressure compartment " refers to a lithostratigraphic body which has an apparently static ($\rho \cdot g \cdot h$) fluid pressure gradient. There are many of these apparently static ($\rho \cdot g \cdot h$) fluid pressure gradients. These are highlighted with a shaded pattern on figure-7.5. Higher permeability rocks have a shorter fluid pressure equilibration time. The apparently static ($\rho \cdot g \cdot h$) fluid pressure gradients may actually be so if the fractures at the minimum work leakpoint are closed at the time of measurement. Pressure gradients across low permeability lithologies have a much longer fluid pressure equilibration time. Thus the apparently static shaded pattern fluid pressure could realistically coexist with the Darcy flow black pattern pressure gradients at the same time.

Darcy flow from any depth to the mudline passes successively through one of the 3 pressure regulators shown on figure-7.5. At any depth, the maximum flow potential of fluids in the earth is the difference between fracture propagation pressure and hydrostatic pressure. The deepest hydrostatic ($\rho \cdot g \cdot h$) fluid pressure gradient on figure-7.5 ends at 12,700 feet and 6100 psi. The Darcy flow pressure gradient below terminates at this depth. A major through cutting sheared fault zone as suggested on figure-7.5 is likely responsible for this extreme depth of hydrostatic pressure.

Four of the fluid pressure gradients portrayed on figure-7.5 terminate at Gaarenstroom's minimum fracture propagation trend line. Many other fracture pressure terminated pore pressure gradients shown on figure-7.3 were eliminated from figure-7.5 for purposes of illustration. The (S_{min}) load on mated fractures determines the highest flow potential of fluids in any tectonic regime basin.

The sheared fracture valves portrayed on figure-7.5 terminate at fluid pressure gradients that are less than fracture propagation pressure. Three of these sheared fracture valves fall within the hydrostatic gradient range delimited by hydrostatic fluid pressure gradient lines. Four of the sheared valves portrayed terminate fluid pressure gradients that are close to the fracture propagation pressure upper limit line. One does not know where the high flow potential water goes from these minimum work valves. One can only say with certainty that the path will end at the lowest intersection with the hydrostatic gradients indicated with hydrostatic fluid pressure gradient lines on figure-7.5.

Two composite fluid flowpaths are portrayed on figure-7.5 that terminate at the upper limit fracture propagation pressure line. A Darcy flowpath at 10,000 feet feeds into the base of a pressure compartment at 9,100 feet. There is a seal relative hydrostatic gradient from there to 8100 feet and 6200 psi which terminates the composite profile. The source of fluid and pressure leading into the Darcy flow segment could well be the mated fracture valve at 10,500 feet and 8500 psi.

The second composite fluid flowpath portrayed terminates at 15,200 feet and 14,500 psi. This flowpath portrays a fracture pressure seal relative hydrostatic gradient of 0.475 psi/ft fed by an underlying Darcy flowpath. Gas is trapped by a caprock at 17,000 feet and 15,000 psi. Water escaping from this pressure compartment feeds the low permeability Darcy flowpath.

Chapter-7.11 Fracture propagation pressure / Pore pressure limit; Conclusions

The concept that fracture propagation pressure is the upper limit to *in situ* pore fluid pressure is fundamental force balance. The concept is in agreement with extensive empirical fracture pressure and pore pressure measurement data. The concept works because fracture aperture and fracture permeability is a positive exponential relationship.

Pore pressure within a fracture opens that fracture at fracture propagation pressure whether the fluid penetrates the formation or not. Pore fluid from below exits through that fracture so long as fluid pressure equals or exceeds fracture propagation pressure. This natural mechanical pressure valve system has Geologic time over which to operate.

Net permeability of mated fractures increases asymptotically as pore pressure approaches fracture propagation pressure (Walsch, 1981). The walls of sheared fractures are uneven and the stress distribution across these fractures is uneven. The uneven-ness causes locally lower minimum stress and relatively lower fluid pressure seal ing capacity. Sheared fracture aperture and permeability are also stress sensitive. Sheared fractures seem to leak at a lower pore pressure than mated fractures.

Intergranular permeability becomes a significant component of net permeability only after the pore pressure sensitive fractures are closed. Fracture permeability is the dominant regulator of subsurface pore pressure profiles. *In situ* fracture propagation pressure is the upper limiting asymptote of fracture permeability. Pore pressure does not exceed this limit for water passing through water wet rocks.

The composite flowpaths portrayed on figure- 7.5 represent pieces of the general Darcy law flow pattern that exists in the subsurface. Fluid pressures vary from a maximum fracture propagation pressure to a minimum hydrostatic pressure. Low intergranular permeability lithologies can form caprocks that can hold fluid pressure up to their fracture propagation pressure. Net permeability within caprocks decays from fracture permeability to intergranular permeability along the fractured flowpath. Intergranular permeability is the flow regulator where pore pressure is significantly below fracture pressure. *In situ* pore pressure profiles are regulated by net permeability. The composite flowpaths on figure-7.5 illustrate a general minimum work force balanced composite Darcy flow regulation relationship.

Knowledge of the *in situ* fracture propagation pressure / pore pressure upper limit relationship can be a tremendous advantage in planning and drilling a particular well. When drilling into a caprock, the maximum pore fluid pressure expected below can be calculated from the position of the minimum work leakpoint. Realistic peak compartment pore fluid pressures can be calculated from the caprock peak fracture pressure. Depth relative pressure gradients in both the caprock and compartment fluid can be determined from local (S_{min}) and fluid ($\rho \cdot g \cdot h$) data. Estimated gas column height from the depth of compartment penetration is the pressure vs. depth scaling factor.

References Cited

Barker, Colin, 1990, "Calculated Volume and Pressure Changes During Thermal Cracking of Oil to Gas in Reservoirs", AAPG Bull, V.74, #8, pp 1254-1261.

Carroll, M M, 1980, "Compaction of Dry or Fluid-filled Porous Materials", Journal of Engineering Mechanics Devision, Proceedings of the American Society of Civil Engineers, Vol. 106, No EM5, Oct 1980 pp969 - 990.

Engelder, T. and Lacazette, A., 1990, "Natural Hydraulic Fracturing", in Barton, N., and Stephansson, O. eds., *Rock Joints*, Rotterdam, A.A. Balkema pp. 35-44.

Gaarenstroom, R.A., R.A.J. Tromp, M.C. deJONG and A.M. Brandenberg, 1993, "Pressures in the Central North Sea: implications for trap integrity and drilling safety", in *Petroleum Geology of Northwest Europe: Proceedings of the 4th Conference* (edited by J.R. Parker) pp 1305-1313.

Grauls, D., 1996, "Minimum principal stress as a control of overpressure in sedimentary basins", in Compaction and Overpressure current research, 8th

Conference on Exploration and Production , Dec 9-10 convenors F. Schneider, & I. Moretti.

Holbrook, P W, 1995a, "The relationship between Porosity, Mineralogy and Effective Stress in Granular Sedimentary Rocks", paper AA in SPWLA 36th Annual Logging Symposium, June 26-29, 1995.

Holbrook, P W, D A Maggiori, & Rodney Hensley, 1995b, "Real-time Pore Pressure and Fracture Pressure Determination in All Sedimentary Lithologies",pp 215 - 222, SPE Formation Evaluation, December 1995

Holbrook, P W, 1996, "The Use of Petrophysical Data for Well Planning, Drilling Safety and Efficiency ", paper X in SPWLA 37th Annual Logging Symposium, June 16-19, 1996.

Kuntze, K.R. & R.P. Steiger, 1992, "Accurate In-Site Stress Measurements During Drilling Operations", SPE 24593, pp. 491-499.

Lorenz, J.C. , NR Warpinski, PT Branagan, AR Sattler, 1989. " Fracture Characteristics and Reservoir Behavior, in Stress-Sensitive Fracture Systems in Flat-Lying Lenticular Formations", JPT June pp615-622.

Nashaat, M., 1998, "Abnormally High Formation Pressure and Seal Impacts on Hydrocarbon Accumulations in the Nile Delta and North Sinai Basins, Egypt", in Law, B.E., G.F. Ulmishek, and V.I. Slavin eds., Abnormal pressures in the hydrocarbon environments: AAPG Memoir 70, p. 161-180.

Walsh, J.B. , 1981, " Effect of Pore Pressure and Confining Pressure on Fracture Permeability", International Journal of Rock Mechanics, Mining Science & Geomechanics, vol 18, pp 429 - 35.

Warpinski, N.R., 1991, "Hydraulic fracturing in Tight, Fissured Media", JPT , February, pp146, - 208.

Chapter-8. HPHT FLUID EXPANSION PORE PRESSURE CALIBRATION, CALCULATION, AND REAL-TIME MWD PORE PRESSURE PREDICTION

Fluid expansion effective stress un-loading requires a very low net permeability seal and a relatively high (>25°C/ km) regional geothermal gradient. Transition from loading-limb to un-loading limb pore pressure calibration depends upon the recognition of the regional seal in each well. A regional un-loading limb stress/strain coefficient is calibrated below the regional sealing surface. The caprock on figure-7.2 is an example of a regional seal.

The depth to fluid expansion un-loading in 16 worldwide basins varies from 2 to 6 kilometers (Ward, 1995). The onset of fluid expansion un-loading occurs between the 90°C and 120°C isotherms in these basins. The occurrence of a regional seal to contain fluid expansion is determined from a regional overburden, high fracture propagation pressure correspondence. These factors control net (intergranular + fracture) permeability that is required to form an effective fluid expansion pressure seal.

In each well within a region, pore pressures above this seal can be determined from a force balanced loading-limb stress/strain relationship. The loading limb stress/strain coefficients sigma max (σ_{max}) and alpha exponent (α) are dependent only on mineralogic constants. These coefficients are composite physical properties of minerals that are independent of region, location and depth. Loading-limb pore fluid pressure could be considerably above hydrostatic before encountering the un-loading pore pressure regime.

The optimum regional un-loading limb *in situ* stress/strain relationship is determined from an appropriate correspondence between measured pore pressures with respect to petrophysically measured strain. A common regional un-loading limb stress/strain exponent (α_{offset}) is determined with respect to the regional fluid expansion seal that is encountered in each well.

Each petrophysical sensor has a different non-linear response that is not readily transformed to porosity through single mineral chart book functions. Petrophysical sensor to porosity transforms depend upon the physical properties of the mineral grains. Each petrophysical sensor has its own borehole environmental problems.

If sensor and mineral specific porosities are different for a given foot, the discrepancy should be resolved before the (strain = 1.0 – porosity) determination is made. The details of the sensor and mineral specific non-linear porosity transforms for resistivity, γ–γ density, and transit-time for water saturated sedimentary rocks are described. Chapter-5. discribes mineralogically sensitive sensor-porosity transforms.

The regional un-loading limb stress/strain exponent (α_{offset}) joins the loading-limb (α) at the regionally extensive seal. This stress/strain fitting approach maintains a mechanically sensible fluid pressure to effective stress/strain proportionality throughout the regionally confined un-loading limb pressure regime. Detailed fluid pressure comparisons in many stacked un-loading limb pressure compartment s indicate that this is usually the case.

Field experience has shown that there is no unique un-loading limb exponent (α_{offset}). Differing regional temperature gradients and complex hydrocarbon richness exert strong influences on *in situ* hydrocarbon cracking. The liquid → gas phase change accompanying complex hydrocarbon cracking is the dominant fluid expansion mechanism.

Mineral grain dissolution can apparently alter the un-loaded sedimentary rocks toward the gravitational loading-limb but not beyond. The gravitational loading-limb stress/strain coefficients (σ_{max} & α) are mineralogically general throughout NFR $\approx$ biaxial basins in the world. The underlying un-loading limb stress/strain exponent (α_{offset}) however, is apparently region specific and needs to be determined empirically as described above.

Chapter-8.1 Pore Pressure and Fracture Pressure derived simultaneously from
Force ⇔ balanced stress/strain physics in the Earth

These are truisms reflecting the governing physics on any stress path. Pore pressure is the fluid load-sharing element in the subsurface. Solid mineral grains bear the remaining load. The fluid scalar pore pressure (P_P) is the difference between the two solid element scalars, average confining load (S_{ave}) and average effective stress (σ_{Ave}), that is ($P_P = S_{ave} - \sigma_{ave}$, Equation-1.2).

Both loading and un-loading pore pressure regimes exhibit a strong dependence on overburden, mineralogy, and *in situ* strain (figure-8.2). One important difference between loading-limb and un-loading limb pressure regimes is their different effective stress/strain coefficients (α) and (α_{offset}).

There are dynamic interactions within the Earth's closed mechanical system that regulate pore fluid pressure in both the loading, and un-loading pore fluid pressure regimes. On a Geologic time scale, fluids are the continuous pore pressure transmission system. The containing solid grain matrix framework and fracture permeability regulate pore pressure profiles (chapter-7. and chapter-8.3).

The over-riding pressure regulation dynamics are that pore pressure at a partially sealing caprock's minimum work leakpoint can be no greater than fracture propagation pressure of the overlying caprock at that leakpoint. Intergranular Darcy flow regulates pore pressures when fluid pressures are below the force balanced fracture propagation pressure limit.

The keys to understanding and quantitatively predicting loading vs. un-loading pore fluid pressure regimes are the subsurface dynamic interactions of these force↔balanced terms. The first key to un-loading limb pore pressure determination is the recognition of a regionally extensive caprock "seal". This is a continuous regional surface between the loading and un-loading limbs in stress/strain space.

The second key requirement is a regionally calibrated un-loading limb stress/strain exponent (α_{offset}) below that seal. These two key factors are inter-dependant. They vary from region to region and must be calibrated from offset wells on that scale.

Chapter-8.2 Stress/strain loading and un-loading hysteresis for granular solids

The loading-limb pore pressure regime contains hydrostatic and dis-equilibrium compaction supra-hydrostatic fluid pressures. The loading-limb stress/strain coefficients sigma max (σ_{max})&(α) are global in nature dependent principally upon mineralogic composition (see chapter-1.0, figure-1.2 and table-1.1). Unloading commences from the peak loading limb point. The un-loading limb stress path follows power-law effective stress/strain exponent (α_{offset}) with respect to the peak loading limb point. This point is shown on figure 8.1 on the following page.

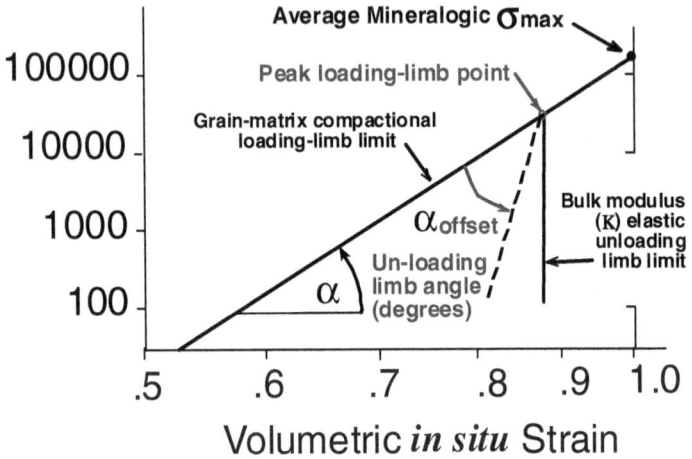

Regional *in situ* loading and un-loading Effective Stress/Strain angular relationships.

Average Mineralogic σ_{max}

Peak loading-limb point

Grain-matrix compactional loading-limb limit

Bulk modulus (K) elastic unloading limb limit

α_{offset}

Un-loading limb angle (degrees)

α

Volumetric *in situ* Strain

The relative *in situ* un-loading limb Effective Stress/Strain response is power-law linear within grain-matrix compactional and Hooke's law elastic limits.

Figure-8.1 is a power-law linear effective stress/strain diagram. The grain-matrix-compactional and elastic mineralogic limits for a porous sedimentary rock are **solid black**. The loading-limb grain-matrix-compactional coefficients (α & σ_{max}) are shown. The grain-matrix-compactional and Hooke's Law elastic stress/strain limits join at the peak loading-limb point (•). The fluid expansion un-loading limb angular offset (α_{offset}) falls between the grain-matrix-compactional loading limb, and the Hooke's Law elastic un-loading limb limits. These coincide at the peak loading-limb point (•).

Each mineralogic end-member has a unique Hooke's Law volumetric elastic stress/strain coefficient (K=bulk modulus). A porous sedimentary rock has a proportionally lower bulk modulus that can be determined from The Extended Elastic Equations that were shown on figure-5.2.

Stress/strain unloading - reloading hysteresis is characteristic of granular solids. The unloading – reloading stress paths approach singularity at geologic strain rates. Hooke's Law volumetric (K=bulk modulus) at the peak loading limb point is the geologic un-loading limb stress/strain limit.

Geologic loading rate volumetric effective stress compaction is power-law proportional to volumetric in situ strain (1.0 - ϕ). The (σ_{max}) mineralogic coefficients are related to the average bond strength and hardness of that mineral's crystalline lattice (see chapter-1.0). The reversible thermal and elastic stress/strain properties of minerals were measured in laboratories decades ago (Carmichael, 1982). The elastic properties of sedimentary rocks are related to and limited by the elastic properties of the minerals of which they are composed.

During grain-matrix loading, sedimentary grains are brought closer together and contact area between grains is increased (see chapter-2.0). The solid element load is borne at grain-grain contacts and through the mineral lattice to the neighboring grains. Under increasing loads elastic energy is accumulated in the mineral lattice in proportion to strain. Elastic strain is a miniscule fraction of total strain, which is dominantly grain-matrix-compactional.

Also during grain-matrix loading, grain contact area is increased irreversibly following a grain-matrix-compactional stress/strain relationship (see chapter-2.0). The limit of grain-matrix-compaction is when all fluid filled porosity is gone and the rock is totally solid. Solidity (1.0 - porosity) is a direct measure of volumetric in situ strain (Chapter-1.0). Compaction of granular solids ends when solidity = 1.0. The volumetric strain of zero porosity rocks involves only thermal and elastic coefficients.

The two grain matrix mechanical limits in the subsurface are 1.) The grain-matrix-compactional loading limb; and 2.) The Hooke's Law elastic un-loading limb. The two crucial elements of regional pore pressure calibration are 1. The location of the peak loading point in the subsurface; and 2. The slope of the geologic un-loading limb effective stress/strain exponent (α_{offset}) below the continuous, regional, peak-loading-limb surface. The caprock in figure-7.2 is representative of this regional surface.

Chapter-8.3 Log and crossplot example showing un-loading limb recognition criteria and stress/strain calibration

The peak-loading-limb to un-loading limb join in a region is a continuous surface. Individual wells within a region define points of this surface. Figure-8.2 is a Gulf Coast example demonstrating how both the peak-loading-limb point and the regional un-loading limb stress/strain exponent (α_{offset}) are related. The entire database represents about 14200 one-foot samples of effective stress and strain data. The upper loading-limb segment is about 9700 feet and the lower un-loading limb segment is about 4500 feet.

Raw in situ Effective Stress/Strain relationships

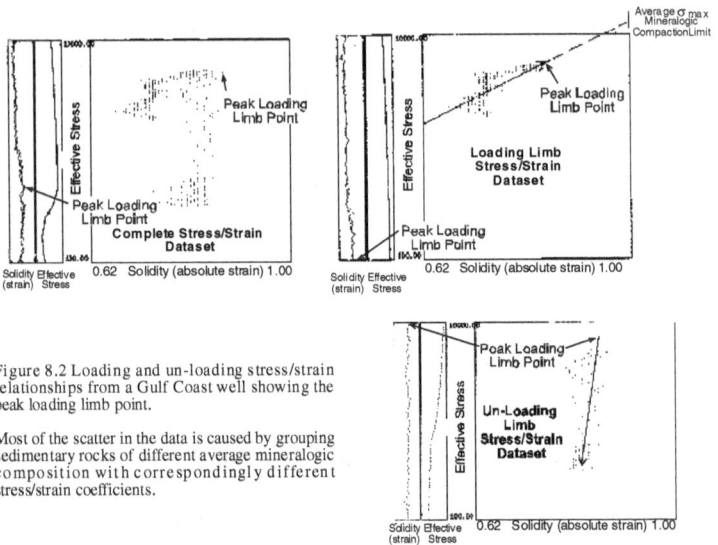

Figure 8.2 Loading and un-loading stress/strain relationships from a Gulf Coast well showing the peak loading limb point.

Most of the scatter in the data is caused by grouping sedimentary rocks of different average mineralogic composition with correspondingly different stress/strain coefficients.

The upper left panel of figure-8.2 shows the complete stress/strain dataset. For clarity, the loading-limb and unloading-limb segments are divided into two separate panels. The loading-limb stress/strain relationship is above. The regional unloading-limb stress/strain relationship joins the loading-limb at the peak loading-limb point. This point is shown on the logs and crossplots of all three panels.

The loading-limb exponent (α) and unloading-limb stress/strain exponent (α$_{offset}$) are distinctly different. The average grain-matrix-compactional loading limb is over 80 degrees and plastic. The average un-loading limb is 89.2 degrees. The un-loading exponent (α$_{offset}$) is very close to the bulk modulus mineralogic elastic stress/strain limit.

There is a slight but measurable increase in porosity due to $in\ situ$ elastic mineral contraction in response to fluid expansion un-loading. Allowable force balance dependent pore fluid pressure measuring relationships must fall between the grain-matrix-compactional loading-limb and elastic unloading-limb limits. The data shown on figure-8.2 represents an extreme but frequently encountered case of fluid expansion un-loading that is near the Hooke's law elastic mineralogic limit.

The location of the peak loading-limb point is clearly discernible on both the stress/strain crossplots and the individual stress and strain logs shown to the side. The peak loading-limb point is the point of highest solidity in this well. This peak solidity point intersects the regionally continuous fluid expansion un-loading pore pressure seal.

Regional seals are known to cross stratigraphic boundaries in the North Sea Central Graben as described by Ward (1995, figure 3). High solidity is the common inter-regional fluid expansion un-loading limb seal recognition criteria in the North Sea, Gulf Coast and Far East. All three regions have approximately biaxial Normal Fault Regime stress fields. The same force balanced physical relationships regulate pore fluid pressure in all three places.

Force balance determines the upper limit pore fluid pressure sealing capacity of all caprocks (see chapter 7). Caprocks above continuous permeable reservoirs can hold no more pore fluid pressure than the minimum principal stress at the caprock's minimum work leakpoint. The minimum work leakpoint is on the same surface as the peak loading-limb point in a particular well that penetrates a particular pressure compartment . A regional caprock seal operates in accordance with solid-fluid force balance and can be recognized by its regionally continuous high solidity. Sealing caprock mechanics are the same in both loading and un-loading pressure regimes.

Chapter-8.4   Far East regional effective stress calibration and real-time pore pressure prediction example.

Six wells in a Far East area were used to establish a regional un-loading limb stress/strain exponent (α_{offset}). Three porosity sensitive sensors, resistivity, bulk density, and P-wave interval transit time were used to determine *in situ* strain (solidity). A baseline normalized gamma-ray sensor reading was used to estimate solid fraction shale volume. A regional average shale grain density profile; and formation water conductivity profile were established.

Mud weight profiles and RFT's were available to calibrate relative and absolute pore fluid pressure. Leakoff tests were available to calibrate fracture propagation pressure. Pore pressure and fracture pressure data from all six wells were evaluated and weighted equally to determine the regional mechanical and petrophysical relationships. Holbrook (1996) describes this calibration procedure up to the point of un-loading limb seal recognition and regional stress/strain exponent (α_{offset}) calibration.

Figure-8.3 is one of the six regional calibration wells used. Mineralogy sensitive raw and normalized gamma-ray readings are shown in track 1. Porosity and mineralogy sensitive bulk density and transit time readings are also shown in track 1. Porosity from "m" variable resistivity is shown in track 1 on the same 0 to 50 PU scale as porosity from bulk density. If the separate sensor→porosity transforms are correct, and there are no borehole related sensor problems, the separate sensor porosity curves should be identical.

Porosity was calculated from resistivity using the second order "Archie" relationship shown on figure-5.1. That porosity transform accounts for the variable mineralogy dependent tortuosity-cementation "m" coefficient. Petrophysical sensor conflicts are resolved. The best calculated porosity enters both the power-law effective stress (σ_{ave}) calculation and the integrated bulk density overburden (S_v) calculations.

The raw Δ_t log is displayed as a darker trace in track 1. The raw Δ_t log tracks the lighter "m" variable Archie porosity log in shales. This correspondence tends to confirm that the input C_w profile used to calculate porosity from resistivity is correct.

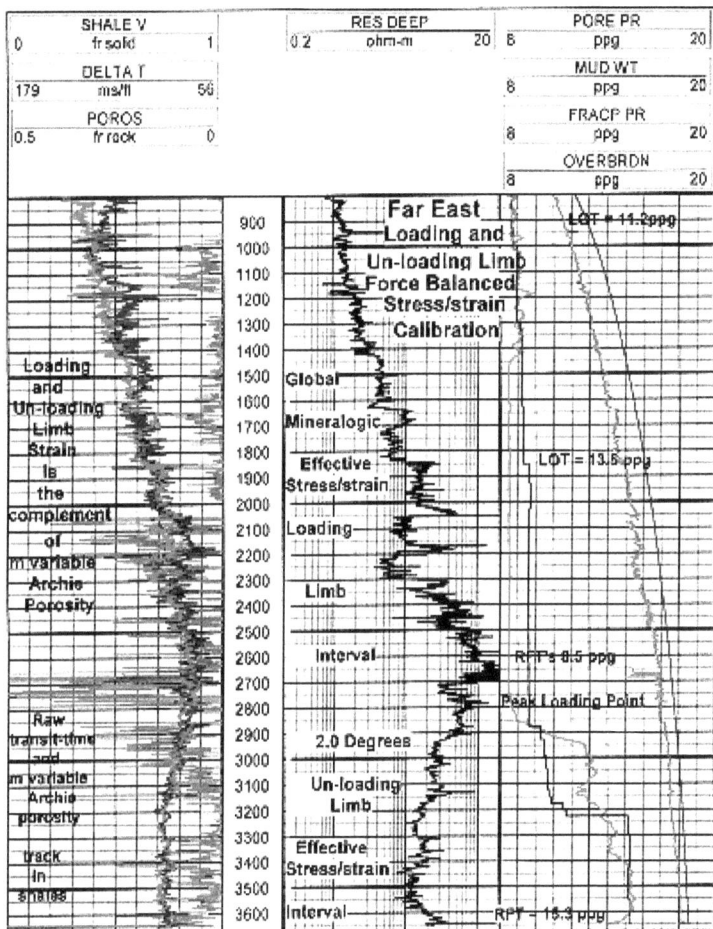

SHALE V		RES DEEP		PORE PR	
0	fr solid 1	0.2	ohm-m 20	8	ppg 20
DELTA T				MUD WT	
179	ms/ft 56			8	ppg 20
POROS				FRACP PR	
0.5	fr rock 0			8	ppg 20
				OVERBRDN	
				8	ppg 20

Figure-8.3. Loading and un-loading limb force balanced calibration well log example. Shale volume and porosity, the complement of strain are displayed in track 1. The deep resistivity data shown in track 2, is converted to porosity using a second order "m" variable Archie relationship which accounts for the mineralogic tortuosity-cementation effect on electrical conductivity. The effective stress/strain loading and un-loading limb intervals both use power-law linear stress/strain coefficient (α_t) or (α_{offset}) to calculate pore fluid pressure. The peak

101

loading point shown on the logs separates the loading and 2.0 degree unloading limb stress/strain regime log segments.

Track 3 displays overburden, fracture propagation pressure, mud weight, and pore fluid pressure in ppg as fluid pressure gradients. The calculated traces are closed-form force balanced. Leakoff tests and RFT's are annotated in track 3 to demonstrate the combined borehole pressure measurements match the closed-form force balanced calibration. The RFT's closely match the force balance calculated pore pressures. The mud weight profile is generally greater than the calculated pore fluid pressure. The green fracture pressure trace exactly matches the upper leakoff test indicating that initial overburden is correct.

Evaluating all the traces, the comparable loading-limb, peak-loading-point, un-loading limb features from figure-8.2 are readily interpreted. The porosity sensitive sensors all indicate a decreasing porosity trend to the peak-loading-point followed by an increasing porosity trend. As in figure-8.3, The peak-loading-limb point is the maximum fracture pressure and minimum calculated porosity of the entire log interval.

The five other calibration wells had these same recognizable features but at different depths. Considering the regional un-loading pressure regime as a whole; a 2.0 degree (α_{offset}) relative un-loading limb stress/strain exponent angle was determined. Force balanced pore pressure, fracture gradient and overburden values on figure 6 were calculated from the 2.0 degree un-loading limb stress/strain exponent (α_{offset}) joined at the peak loading point.

Figure-8.3 shows a long interval of increasingly under-balanced drilling from 3000 to 3500 meters. This was also true in several other calibration wells. Drilling mud gas did not provide adequate warning of under-balanced drilling in these wells or the well that was logged MWD. Some of the calibration wells were shut in to control un-predicted pore pressure . One of these wells required hole abandonment and side tracking to reach their drilling objective.

Chapter-8.5   Pore Pressure and Fracture Gradient Prediction from MWD data

Figure-8.4 shows the final log generated from combined wireline and MWD petrophysical data. No adjustments were made to any of the regionally derived constants or control profiles for the entire duration of this well. This well also shows the same upper Global Mineralogic Stress/Strain Loading-limb interval.

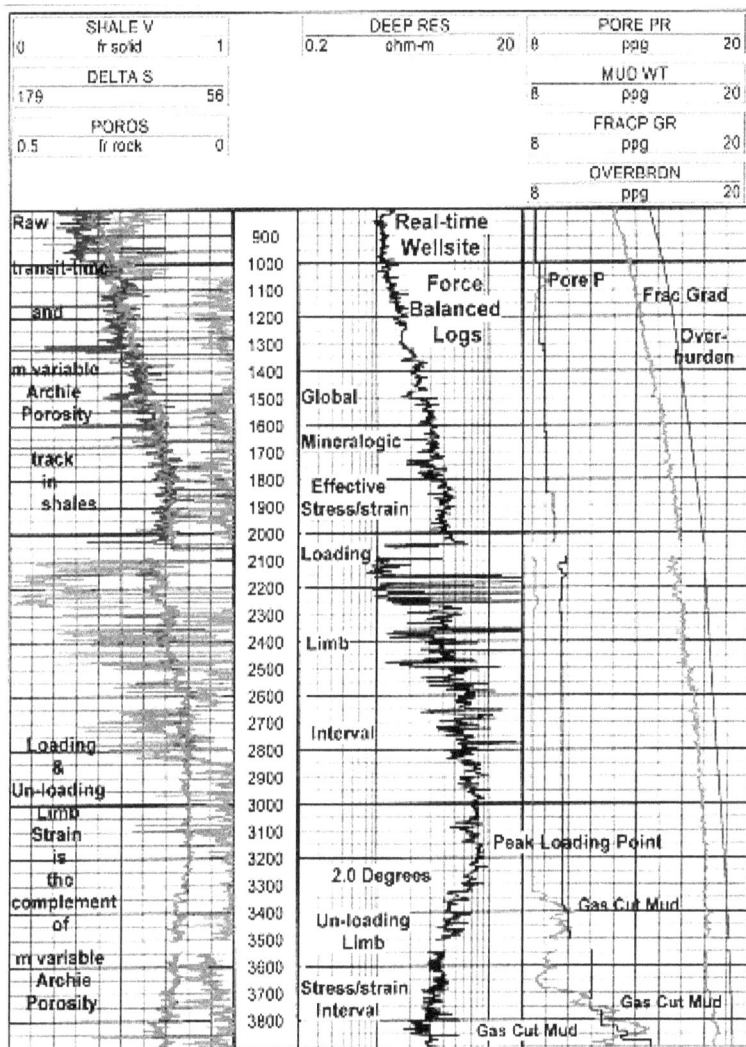

SHALE V			DEEP RES			PORE PR		
0	fr solid	1	0.2	ohm-m	20	0	ppg	20
DELTA S						MUD WT		
179		56				8	ppg	20
POROS						FRACP GR		
0.5	fr rock	0				8	ppg	20
						OVERBRDN		
						8	ppg	20

Figure-8.4. Real-time Force balanced logs in the same format as figure-8.3. All petrophysical calibration constants are identical to the regional calibration used for figure-8.3.

The only operator intervention applied during the entire well was to switch to the regionally calibrated 2.0 degree (α_{offset}) un-loading limb stress/strain coefficient at the peak loading-limb point that is annotated on

103

figure-8.4. The same 2.0 degree relative un-loading limb stress/strain exponent (α_{offset}) was used below the peak loading-limb point.

The real-time pore pressure log indicated under-balanced mud weight at 3380M. The first gas cut mud incidents occurred at 3400 M. The kill weight of 10.5 ppg was in agreement with that predicted by the real-time pore pressure log.

Mud weight was raised again to 10.7 ppg at 3480M in preparation for setting casing. The primary objective of setting casing in the overpressured interval was achieved. It is important to note that the onset of elevated pore pressure occurred 200 meters below the peak loading-limb point and 200 meters below the resistivity reversal on the raw resistivity log. Based upon the above mentioned mud cut and kill weights the regional 2.0 degree un-loading limb stress/strain calibration was correct.

Mud weight was raised to 11.6 ppg after casing was set. A sharp pore pressure increase to 11.6 ppg occurred at 3725M. Under-balanced mud weight was predicted at 3750M. This preceded gas cut mud and mud weight was raised to 12.3 ppg that temporarily controlled the mud gas cut.

Heavy gas mud cut occurred again at 3800M. The eventual kill weight for this gas cut mud was 14.8 ppg in agreement with the real-time pore pressure . The flow and kill mud weights bracketed the force balance calculated pore pressure to 0.2 ppg accuracy. The operator had never before witnessed this level of accuracy and was very pleased.

Chapter-8.6   Regional Un-loading limb calibration & prediction; Conclusions

Pore pressure accuracy resulted from the application of a composite loading limb and un-loading limb mechanical systems method. The method corresponds to a regional loading-limb regime over a fluid expansion un-loading limb pressure regime that was related to petrophysical data (figure-8.1). The predictions were made through a regionally calibrated closed-form mechanical system. Both wireline measured and MWD petrophysical data were used as strain measurement input to the mechanical system.

Drilling efficiency benefits were, 1.) The gas-cut mud incidents were circulated out without well shut in. 2.) The operator was able to safely reach his drilling objective with one less casing string by setting casing in the upper portion of the un-loading limb pressure transition zone. Safety and cost objectives were met by applying force balanced physics in the calculation of pore pressure.

Using force balanced stress/strain mechanical relationships to directly determine pore pressure is a sound drilling engineering choice. Pore pressure is mechanically related to porosity, mineralogy and density that can be estimated from several petrophysical sensors in the closed mathematical form (see figure-6.1). The calculation procedure applied incorporates _in situ_ measured rock grain-matrix stress/strain relationships (table-1.1) with Newtonian physics that is mathematically related to absolute _in situ_ strain through equation-1.1 is described in chapter-1.0 and chapter-6. This is the only pore pressure method that involves constitutive mechanical systems.

Loading and un-loading limb pore pressure regimes have distinct separate stress/strain exponents (α) and (α_{offset}). These relationships were shown on figure-8.1, and figure-8.2. The effective stress/strain regime join in the subsurface is a continuous high solidity caprock that partially seals an expanding fluid phase pore pressure . Figure-7.2 is an exemplary of this. Pore pressures in both the loading and un-loading fluid pressure regimes are treated with mechanically sensible mineralogic stress/strain relationships.

Closed-form force balance comprehensively accounts for both fluid and solid borne subsurface loads. Pore pressure below the peak-loading-limb surface is limiting by the fracture propagation pressure at the Minimum Work Leakpoint of the sealing caprock (see figure-7.2). Pore pressure in both loading and un-loading regimes has been determined from regionally calibrated effective stress loading and un-loading stress/strain mathematical functions. This is a unique mechanical systems approach to a very difficult pore pressure prediction problem.

References cited

Holbrook, P.W., 1999c, "The primary controls over sediment compaction and their force balanced relationship to pore pressure". **Pressure Regimes in Sedimentary Basins and their Prediction** AAPG Special Volume.

Holbrook, P.W., 1998a, "Physical explanation of the closed form mineralogic force balanced stress/strain relationships in the Earth's sedimentary crust." Presented at **Overpressures in Petroleum Exploration,** the European International Pore Pressure Conference, April 7-8 Pau, France. Published in Elf special volume 1999.

Holbrook, P W, 1998b, "The universal fracture gradient/pore pressure force balance upper limit relationship which regulates pore pressure profiles in the subsurface", in AADE Industry Forum on **Pressure Regimes in Sedimentary Basins and their Prediction,** Sept 1-4, 1998 Del Lago, Texas

Holbrook, P W, 1996, "The Use of Petrophysical Data for Well Planning, Drilling Safety and Efficiency ", paper X in SPWLA 37th Annual Logging Symposium, June 16-19, 1996.

Holbrook, P W, 1995a, "The relationship between Porosity, Mineralogy and Effective Stress in Granular Sedimentary Rocks", paper AA in SPWLA 36th Annual Logging Symposium, June 26-29, 1995.

Holbrook, P W, D A Maggiori, & Rodney Hensley, 1995b, "Real-time Pore Pressure and Fracture Pressure Determination in All Sedimentary Lithologies", pp 215 - 222, SPE Formation Evaluation, December 1995

Ward, C.D., 1995. , Coghill, K. and Broussard, M.D. " The Application of Petrophysical data to improve pore pressure and fracture pressure determination in North Sea Central Graben HPHT wells". SPE paper 28297 presented at 1994 SPE Annual Technical Conference and Exhibition, New Orleans, Sept 25-28.

Chapter-9. EMPIRICAL "NORMAL COMPACTION vs. DEPTH TREND" PORE PRESSURE METHODS.

Empirical depth trend pore pressure and fracture pressure prediction methods have existed for 35 years. There are more than 250 of these depth trend methods. These methods all relate a measured parameter that is not **strain**; to depth that is not a **stress** variable. This huge diversity of methods, combined with the lack of a mechanical systems model makes empirical method selection the principal focus of today's pore pressure prediction specialists.

These individual sensor vs. depth methods predict pore pressure successfully when the specialist operator causes a partial intersections of the empirical parameters with the earth's mechanical **stress/strain** system.

All the indirect methods depend on some sort of empirical curve fits. Stress, strain , force balance and corresponding dimension units are not explicitly used. There are no mechanical boundary conditions. Units do not correspond in these Empirical pore pressure methods. Non-dimensional empirical coefficients are used to bridge the dimensional gaps as well as the gaps in mechanical understanding.

Table-9.1 Shows a representative set of empirical pore pressure methods. Each major table heading are; 1. The author and date of the method publication. 2. The measured parameters being correlated; 3. The means of establishing a "Normal" compaction trend (**Pn**), and 4. The means of establishing an "excess" pore pressure (**ΔP**) calibration.

Chapter-9.1 Explanation of Indirect Empirical Pore Pressure Prediction Methods

Table-9.1 is a representative lists of **Empirical pore pressure methods**. The authors and their publication dates are listed alphabetically by row. One could get more information on any of these references by looking up that author in that year.

The indirect empirical methods are first categorized according to the **Measured Parameter** (surrogate strain) that is correlated to depth (surrogate stress) from which pore pressure is calculated.

Empirical (Pp = Pn + ΔP) — Pore Pressure — Prediction Techniques

Authors	Measured Parameters (ROP Res Dens Vel -ray temp)						"Normal" Compaction Trend (Pn)		HOW	Excess Pressure Calibration (DP)		
	ROP	Res	Dens	Vel	-ray	temp	Empirical	Prescribed		Equivalent Depth	Empirical	1 Overlay
Belotti, Gerard (1976)	ROP	Res	Dens	Vel	-ray	temp	Empirical	Prescribed		Equivalent Depth	Linear + Offset	1 Overlay
* Belotti, Giacca (1978)	ROP		Dens				Empirical		Overburden		Calculated Equation	
Belotti, Giacca (1978)	ROP						Empirical		Linear Offset		Seismic Inversion	
Bilgei, Ademero (1982)	ROP						Empirical					
* Bolt, D.B. Jr. (1976)	ROP	Res	Dens		-ray		Empirical		Linear			Matching
Boatman, W.A. (1976)	ROP	Res	Dens				Empirical		Linear Fit		Non-Linear Deviation	
* Burgoyne, Young (1974)	ROP	Res	Dens				Empirical		Multivariate		Regional Equation	
* Burgoyne, Rizer &	ROP	Res	Dens				Empirical					
Myers (1987)	ROP	Res	Dens				Empirical				Linear	
Combs (1968)	ROP	Res	Dens				Empirical		Linear Fit		Pressure Charts	
* Eaton, B.A. (1972)	ROP	Res	Dens				Empirical		Overburden		Resistivity Ratio	
Fertl, Chilingarian (1977)	ROP	Res					Empirical		Linear		Relative Deviation	
Fertl, Illavia (1977)	ROP	Res					Empirical				Overlay	
Fontenot, Berry (1975)	ROP	Res					Empirical		Linear		Equation	
* Foster, Whalen (1965)	ROP	Res					Empirical		Formation Factor	Equivalent Depth	Shales Only	
Gill, J.A. (1983)	ROP	Res					Empirical		Linear+Offset		Shales Chart Overlay + Only	
* Gill, J.A. (1983) (Unpublished)		Res					Empirical		Linear+Offset		Non - Linear / Offset Matching	
Greene, K. (1978)	ROP	Res					Empirical				Shales Only	
Hamour, Muller (1984)	ROP	Res					Empirical		Prescribed	Equivalent Depth	Shales Pressure Limits	
* Hottman, Johnston (1965)	ROP	Res					Empirical		Linear Fit		Non-Linear Deviation	
** Jorden, Shelley (1966)		Res					Empirical				Match	
* Mathews, Kelley (1967)		Res					Empirical				Non-Linear Deviation	
McClesky, R.W. (1968)	ROP	Res					Empirical				Matching	
McKee, Pakington (1974)	ROP	Res	Dens				Empirical				Shales Only	
Moore, E.L. (1974)	ROP						Empirical					
Prentice, C.M. (1980)	ROP						Empirical		Linear+Offset			
Prentice, C.M. (1980)	ROP						Empirical		Multivariate			
Simpson, N.A. (1984)	ROP						Empirical				Non-Linear from Trend	
Snyder, Suman (1978)	ROP		Dens				Empirical					
* Stein, N. (1985)			Dens					Prescribed	Niagara			
Vidrine, Benit (1967)	ROP						Empirical		Linear Fit			
Wallace (1965)					-ray		Empirical		Linear			
Zoeller, W.A. (1978)	ROP		Dens				Empirical					
Zoeller, W.A. (1984)	ROP		Dens				Empirical					

The six raw measured parameters used are abbreviated as; **ROP.** = drill rate, **Res** = resistivity, **Dens** = density, **Vel** = velocity, **γ -ray** = Gamma-Ray, and **Temp** = temperature. If the **Author** used one of these sensor measurements as a (surrogate strain) variable the sensor is listed on that author's row.

The second category in the table describes **HOW** the author establishes his "**Normal**" Compaction Trend (Pn) from the **Measured Parameter** vs. depth compactional relationship. If the described method is purely **Empirical** , that is shown on the author row in the his "**Normal**" **Compaction Trend (Pn)** column. If the depth correlation is described specifically **Prescribed** that appears in the row and column. Where there is a clear indication as the author's approach to the compactional relationship, it is summarized in a few words under the **HOW** column.

Pore pressure is often inferred to be hydrostatic pressure in the "normally compacted" (**Pn**) **Measured parameter** calibration depth range. In any given combination of methods, there are as many "normally compacted" vs. depth trends as there are sensors. Lacking a physical basis for the **Measured parameter** correlation, operator expertise and judgement is used to decide which sensor works best under which geologic conditions. The choice depends on some composite of petrophysical, and mechanical set of beliefs that can probably not be specified.

Next, there are three categories describing How the Excess Pressure Calibration (**ΔP**) is made. In all cases (**ΔP**) is a term which is added to (**Pn**) to yield (**Pp**) pore pressure . The most fundamental problem with all these methods is that a solid rock attribute is indirectly measured and related to a fluid pressure. Neither the solid-fluid mechanical inferences of what is being measured are related nor do the units of measurement correspond.

If the pore pressure is somehow calculated from the identical sensor reading on the "**Normal**" **Compaction Trend (Pn)** empirical relationship, the row and column indicates that one of the Equivalent Depth Methods was used. If the method is essentially a re-scaling of the difference from the extended (**Pn**) trend to match a **measured pore pressure** (Pp), some brief descriptive comments are made in the **EMPIRICAL** column. If the described method involves some sort of transparent overlay, a brief description is given under in the rightmost **OVERLAY** column.

A "Normal" Compaction vs. Depth Trend (Pn) by necessity incorporates a variable Overburden Gradient in the compaction trend. For Equivalent Depth Methods to be physically representative the Overburden Gradients at the two depths being compared should be the same. Due to the effect of compaction on density and Overburden Gradient this is certainly not true. There are many Equivalent Depth Methods despite the use of this false assumptions.

Although all the listed Empirical methods in the table depend on an earth stress/strain relationship, none of these Empirical methods involves a measure of stress or strain. Fertl and Chilingarian, (1987) graphically summarized many of these indirect-empirical-depth-trend-methods.

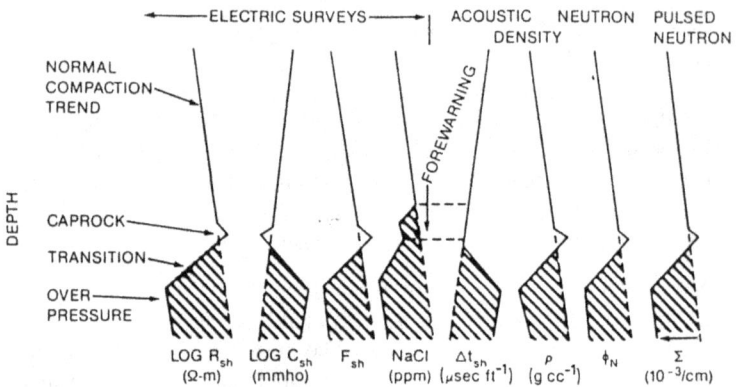

Figure-9.1 Schematic response of well logging parameters to normal and overpressure environments. C_{sh} = shale conductivity(mmho); R_{sh} resistivity (Ωm), resistivity log; F_{sh} = Formation resistivity Factor, dimensionless; Δt_{sh} = acoustic log; r = bulk density (g/cm3, density log; Φ_N = porosity index (%) neutron log; Σ_{sh} = neutron capture cross-section of shale (10-3), PNC log. After Fertl and Chilingarian, 1987, fig. 2.)

Figure-9.1 shows some representative petrophysical sensor vs. depth trends. All indicate an end-of-normal compaction depth. A caprock is shown on each trend. Though useful for trend identification, many pore pressure increases occur gradually without evidence of a caprock. A gradual onset of "excess pore pressure" is a serious problem for an operator attempting to apply a "Normal compaction trend" method.

The "under-compaction" fluid pressurization mechanism often begins gradually following a loading-limb stress/strain relationship. The "fluid-expansion" fluid pressurization mechanism often begins with a caprock but follows a regional unloading limb stress/strain relationship within an over-pressured zone.

It is unclear how these necessarily different (**Pn**) vs. (**ΔP**) pore pressure trend scaling effects should be captured in the slopes of the "overpressured" line segments on figure-9.1. The separation from the projected "Normal compaction trend" is the overpressure scaling factor. Distance below the base of the sloping "Normal compaction trend" is now an important factor. This does not make sense mechanically, but can be measured and scaled empirically to the trend line extension.

Chapter-9.2 Depth Functions equate to Shales Only Pore pressure Methods

Empirical pore pressure methods that take the step of converting a **Measured Parameter** into a depth function are thereby limited to shales only application. Actually all of the techniques listed on the Empirical Mothods table-9.1 apply to shales only.

There is no *a priori* way to predict lithology at a given depth. Shales, limestones and quartz grainstones all have different porosity transforms (see Chapter-5.). Different lithologies also have different compactional stress/strain relationships (see figure-1.2 , table-1.1 , and figure-2.7).

The shales only limitation forces the method operator into the step of editing a continuous petrophysical logs into log segments that include only end-member shales. There is considerable work required to execute this step and one is left with much less data than was originally available.

The shales only problem goes further. The most important goal is to predict overpressure in permeable lithologies that might flow if they were penetrated with an underbalanced borehole fluid pressure. The formation pore pressure in quartz grainstones is often significantly different from the surrounding shales. The shales only limitation forces this risk that can lead to unexpected well flows and blowouts if borehole fluid pressure is underbalanced. Loss of drilling fluid and formation damage can occur if borehole fluid pressure is overbalanced. These risks are an undesirable but necessary consequence of the shales only limitation.

Chapter-9.3 Alternative Grain-Matrix-Compactional constitutive relationship

All of the methods listed on the Empirical Methods table-9.1 have the shales only limitation. They also avoid the obvious solution that granular compaction, pore pressure, and fracture pressure can be formulated as a force balanced in situ stress / strain relationship. All of the following are unique features of this grain-matrix-compactional mechanical systems approach to the pore pressure prediction problem.

1. The mechanical system has a closed-form, dimensionally correct, physical boundary condition; v:H:h confining loads (S) – v:H:h effective stresses = Pore pressure.
2. Bulk density and Overburden Gradient are calculated from input petrophysical and geometric data.
3. Porosity and its complement solidity are calculated from mineralogic and NaCl Brine conductivity sensitive petrophysical transforms.
4. This mechanical system calculates the full range of effective stresses for two lithostratigraphic continua; Quartz grainstone-claystone, and limestone-claystone.
5. All effective stresses and strain are power-law related to grain-matrix-compactional strain in this closed-form mechanical system.
6. Units of depth, density, pressure, pressure gradients, and porosity match in this mechanical system. The in situ coefficients have been verified empirically.

Overburden that is calculated within the mechanical system is free of *a priori* depth trend bias. It depends only on a top-of-petrophysical-data constant plus values calculated from the input petrophysical data.

Chapter-9.4 Benefits of Earth Mechanical Systems for Pore Pressure Determination.

Most of the difficulties associated with empirical depth trend methods can be avoided.

1.) Bulk density – overburden – porosity correspondence is a dimensional mechanical system attribute that is executed through Newton's law instead of an operator adjustable decision. Equation-1.1 is that gravitational relationship.

2.) The earth mechanical system is by definition directly strain (1.0 - φ) dependent.

3.) All the separate sensor calibration difficulties demonstrated on figure-9.1 are resolved by first using mineralogically sensitive, calculated porosities. (See Chapter-5.)

4.) Closed-form force balance governs pore pressure. When applied, this mechanical system boundary condition physically constrains pore pressure calculations.

5.) The fracture pressure/pore pressure limit (See Chapter-7.) is an explicit solid-fluid mechanical system pore pressure regulator. It is general to all tectonic regimes. All empirical pore pressure methodologies that lack Effective Stress Theorem boundary conditions also lack the means to apply this real mechanical systems regulator.

When applied, mechanical system constraints direct operators toward more meaningful decisions. Pore pressure interpretation problems should be solved more quickly and easily by focusing only on those relationships that are not already mechanical system requirements.

References cited

Fertl, W. H. and Chilingarian, G.V., 1987, "Abnormal formation pressures and their detection by pulsed neutron capture logs", Journal of Petroleum Science Engineering, 1(1): 23-38.

Authors listed in table-9.1

Chapter-10. UNDERSTANDING THE EARTH MECHANICAL SYSTEMS

Earth mechanical systems are a new conceptual approach to subsurface geology and geologic engineering. Porosity, mineralogy and strain are characterized as a force-balanced response to loads applied in the subsurface. The solutions to many present and future engineering problems are directly dependent on the earth's mechanical systems.

The earth is a mechanical system much like a lever. The earth is composed of minerals and fluid that both generate and bear subsurface loads. The earth is a closed mechanical system that is encased in similar earth material. In between fault movements and in between faults, effective stress has almost infinite time to equilibrate to *in situ* strain.

The universal Effective Stress Theorem boundary condition and a definition of absolute *in situ* strain have been combined with Newton's law and Hooke's law and Pascal's principle to define two compositionally related earth mechanical systems.

Recognition and understanding of the earth mechanical system transforms pore pressure from an empirical independent variable to a mathematically dependent variable within a closed form. Pore pressure, fracture pressure, and overburden are earth mechanical system mathematical descriptors. Porosity is a binary mineral – fluid partitioning coefficient and a measure of grain matrix *in situ* strain.

Using the earth mechanical system as a load descriptor provides related mathematical answers to important engineering questions. Pore pressure, fracture pressure, and overburden are related to material properties in several mechanical systems domains. These material properties feed into Hooke's law, Newton's law (equation-1.1) and an earth grain-matrix-compactional mechanical system (see chapter-6. & figure-6.1).

Chapter-10.1 The Future of Earth Mechanical Systems

Earth mechanical systems are broadly applicable to well-planning, drilling, casing, completion, geotechnical and reservoir engineering. Each of these subsurface engineering applications uses elastic, density, and porosity coefficients. The confining v:H:h loads, effective stresses, and pore pressure are also used. These parameters are physically linked to each other through the earth's constitutive mechanical systems.

The classical physics of closed-form force balance, Newtonian gravitation and Hooke's law apply directly and simultaneously in the earth's sedimentary crust. The earth mechanical systems necessarily respond to the governing physics whether that is understood by the observer or not.

The confusion generated by the proliferation of empirical pore pressure methods has left an impression that underlying earth mechanical systems are unknown and perhaps un-knowable within the very complex earth. The 250+ empirical pore pressure methods correspond to measurement correlation's without any direct attempt toward mechanical systems representation. The different empirical pore pressure methods produce different results furthering the un-knowable earth perception and the confusion.

Better statistical correlations support a unifying hypothesis wherein forces are balanced. The hypothesis further holds that the earth is a classical mechanical system wherein effective stress is everywhere proportional to in situ strain.

Chapter-10.2 The promise of Earth Constitutive Mechanical Systems.

Mechanical systems hypotheses are the basis for many physical sciences. It is common practice in science to relate theory to measurements. Statistics are used to validate or invalidate predictive systems models. The hypothesis that the earth can be represented by one or several mechanical systems domains is certainly reasonable.

The chapters and figures in this book show correlations of physical parameters that are dimensionally related to each other and to several discrete earth mechanical systems models. The correlations to mechanical systems models shown are as-good-as or better than the correlations used by empirical pore pressure methods. Correct dimensionality and use of only material properties coefficients are two additional advantages of the earth mechanical systems models presented above.

The hardware aspects of subsurface engineering are already practiced with respect to man-made mechanical systems. The empirical unknowns have been in the perceived to be complex earth. Perceived complexity is actually generated by fundamentally non-causal empirical inferences. One cannot reach mechanical systems conclusions through non-causal methods that are not related to mechanical systems boundary conditions.

As shown above the earth can be characterized as a linked set of mechanical systems. An earth surface analogy could be the linkage of a pendulum to a lever to a wheel that we know of as a clock. We understand these separate mechanical systems above the earth's surface and can make them work together. The same can be done beneath the earth starting from the earth mechanical systems described in this book.

The twenty-four parameters listed in table-6.1 through table-6.5 are related through combination of the Extended Elastic Equations with Grain-matrix-compactional mechanical system models. The evidence mentioned above suggests that these two mechanical systems models are accurate. The earth mechanical systems described relate physical laws to measurable mineral and fluid properties.

Seismic and borehole elastic waves propagate according to Hooke's law. There are several mechanical systems solutions to the problem of predicting subsurface pore pressure from seismic data. Porosity is also predictable from seismic data through petrophysical-mechanical systems linkages. This direct mechanical approach may be realized in the future.

The prediction of in situ loads and stresses within thrust fault regimes is conceivable through the same closed-form mechanical systems approach. A thrust fault block is a mechanical unit in the same sense as a normal fault bounded block or a strike-slip faulte bounded block. The in situ stress ratios are close to even number relationships within the former two fault regimes. There are enough stress and strain measurements to determine the general Load State in a thrust fault block as well.

Borehole stability is dependent on porosity, mineralogy and hole angle within a borehole radial stress field. Much is known about both borehole radial and earth v:H:h stress fields. They can be combined as mechanical systems to form a new predictive basis for borehole stability.

There is every reason to believe that mechanical systems approaches to these problems will work as well as or better than empirical approaches. Known boundary conditions such as the effective stress theorem, equation-1.2 and Newtonian gravitational load, equation-1.1 will always apply. The porosity fluid-solid partitioning coefficient and its complement (solidity=strain) should be involved no matter how a mechanically significant solution is formulated. Fitting in situ measurements to a mechanical systems model should be one approach, if not the first approach that should be tried.

Pore Pressure through Earth Mechanical Systems
Cross Referenced Glossary of Technical Terms

Alpha - the Greek letter α corresponding to the power law exponent in the First Fundamental force balanced effective stress / solidity in situ compactional relationship for granular solids. Chapter-1.0 and figure-1.2 explain alpha's relationships in greater detail.

Anhydrite - $CaSO_4$, common evaporite mineral, hardness = 3.5, solubility @ STP = 1000 ppm . The second mineral to precipitate from the evaporation of average seawater. Intermediate plastic compressibility evaporite mineral. Table-1.1 and figure-1.2 show anhydrite's mechanical properties with respect to other minerals.

Biaxial - A stress state where two of the three orthogonal stresses are equal. In a Normal Fault Regime Basin, the maximum principal stress is vertical. The two horizontal principal stresses are usually very close approximating a biaxial stress field. Most Laboratory Rock Mechanics Systems are biaxial wherein maximum principal stress is axial relative to a jacketed cylindrical rock sample. The article, "The use of petrophysical data for well planning, drilling safety and efficiency", by Holbrook (1996) shows the comparable force balanced data input control for the laboratory and In situ Rock Mechanics Systems both of which are biaxial. Figure-4.1b shows the biaxial stress field relationship in the Gulf Coast.

Borehole fluid pressure - The hydrostatic force / unit area at a given depth in a borehole. The measurement will be different depending on whether or not the borehole fluid is circulating (ECD) of static.

Borehole fluid pressure gradient - The average density of borehole fluid in the annulus of a borehole including entrained cuttings. Typical units are grams/cubic centimeter or pounds/gallon. syn. Mud weight.

Calcite - $CaCO_3$, Hardness = 3, Solubility @STP= 140ppm. A very common mineral; The first mineral to precipitate from the evaporation of average seawater. The major constituent of limestones, and marble. A common minor constituent in quartz sandstones and shales. Calcite effervesces in dilute acid solutions. Highest plastic compressibility evaporite mineral. Figure-2.8 shows a well log demonstrating pore pressure in mixed calcite-clay lithostratigraphic sequences.

Capillary pressure - The additional pressure required to force a non-wetting phase fluid through a capillary size pore. It is the difference in pressure between the concave and convex sides of the curved immiscible fluid interface multiplied by the interfacial tension between the two fluids.

Capillary pressure increases markedly with decreasing pore throat size and increasing interfacial tension. Chapter-7. provides a general explanation of the caprock fracture permeability relationships to hydrocarbon trapping and pore pressure profiles.

Caprock - Low intergranular permeability lithotype overlying a continuous pressure compartment . Figure-7.2 shows the caprock's geometric relationship to pressure compartments. Chapter-7. provides a general explanation of the caprock relationship to pore pressure profiles.

Cement - Minerals precipitated in the inter-granular pore space of a supporting grain matrix. Cement is precipitated preferentially in natural fissures or fault planes that conduct hotter more concentrated fluids from below toward the surface.

Figure-2.2 a shows the relationship between pressure solution dissolves and pressure solution re-precipitated cement where solid mass has been conserved. Mass balance calculations indicate that in most cases the cement found in sedimentary rocks away from high flow conduits is locally derived from the surrounding particle grains.

Clay - aluminosilicate composed of alternating silica tetrahedral and aluminum octahedral layers. Sedimentary clays are typified by fine (less than 5 micron) particle size. Clay particle aspect ratios are from 50:1 to 200:1. Clays are internally charge balanced but have consistent surface charges that give them similar mechanical properties. The surfaces of all clays are negatively charged and the edges positively charged. Clay minerals have hardness between 2.5 and 4. The solubility of sedimentary clay minerals @STP is low from 10 to100ppm. Clay minerals are the dominant constituent of shale. Figure-2.1 shows a clay structure and Figure-2.2 c shows a clay mineral interlamellar pore space.

Compaction - The dimensional reduction of a solid body by an applied force. The dimensional change can be anywhere between entirely reversible elastic or entirely irreversible plastic with respect to applied force. See chapter-2.0 for a detailed explanation of *in situ* grain-matix compaction.

Confining Stress - The external load applied to a body of matter. Confining stresses are vectorial v:H:h with respect to the earth's surface coordinate system. Overburden is the vertical confining stress in the earth. The average confining load is volumetric. See the effective stress theorem, and equation-1.2

Continuous Pressure Compartment - A moderate to high permeability lithostratigraphic body that has equilibrated to seal relative hydrostatic pressure over Geologic time. See Figure-7.2 in Chapter-7.

Critical porosity – The break in slope from an approximately linear velocity vs. porosity relationship in sedimentary rocks (Nur et al, 1991). The slope break is approximately the highest porosity at which a sediment - fluid mixture will transmit a shear wave.

This "critical" porosity occurs near the transition from a grain supported solid to a solid - fluid slurry. Critical porosity corresponds approximately to the Wood equation limit in the Extended Elastic Equations. See figure-5.2. , Chapter-5. and Chapter-6. for other related mechanical systems information. The ASTM liquid limit test measures this porosity as a volume percent water content of a saturated soil at the transition state.

Darcy's law - The relationship for non-turbulent single phase fluid flow through permeable granular material that governs natural subsurface fluid flow volume to relative fluid pressure and velocity to permeability and flow area over a certain distance. Net premeability is often dominated by fracture permeability as documented in chapter-7.

Density - Mass / unit volume. Grams / cubic centimeters in metric units. Density is a controlling earth material property in the Extended Elastic Equations and the grain-matrix-compactional mechanical systems.

Disequilibrium Compaction - The excess fluid pressure generating mechanism that results from increased geologic overburden loading which occurs at a rate faster than which fluid can escape the intergranular pore space of a compacting sedimentary rock. Disequilibrium compaction is generally considered to be the dominant pressure generating mechanism in rapidly subsiding Normal Fault Regime basins.

Drill Stem Test - A pressure measurement made in a cased borehole. DST's are routinely made at the start of production to estimate and forecast production rates.

Effective Stress - The fraction of confining stress that is borne by the solid fraction of a body of matter. Effective stress is generally applicable to porous granular solids.

Effective Stress Theorem - General volumetric force balance definition stating that the external confining stress on a body of matter is borne by the sum of the Pore fluid pressure plus the average Effective stress (Carroll, M.M., 1980).

Elastic - A reversible stress/strain relationship wherein the sample being tested returns to its original dimension when the stress is removed. Hooke's law (1640) holds that strain is proportional to stress. A set of elastic moduli describes the inter-relationships of elastic properties and density for isotropic solids. See Chapter-5. and Chapter-6., as well as Table-6.1 through Table-6.5. See figure-5.2 that shows the Extended Elastic Equations mechanical system domain.

Empirical pore pressure methods - Non-physical pore pressure interpretation assistance techniques that involve empirical relationships in two different but mechanically inseparable pore pressure regimes.

A "Normal compaction" **Pn** empirical relationship is determined by plotting a measured parameter vs. depth in a zone that is believed to be at hydrostatic fluid pressure in communication with the surface. An "Excess pressure" (ΔP) empirical relationship is determined by plotting a difference in the previous measured parameter vs. depth in an "overpressured" zone that is believed to be above hydrostatic fluid pressure with partially blocked fluid communication to the surface. Figure-9.1 summarizes these empirical methods graphically. See Table-9.1 for a list of these methods.

All these methods rely upon a direct comparison between a raw petrophysical measurement which changes in the same lithology with depth. Implicitly the change in that measurement is related to compactional strain and pore pressure.

The Empirical pore pressure methods operate only through petrophysical measurement comparison in shales only. The operator has the responsibility of determining whether the shales that are being measured at both comparison depths have similar compactional responses. Refer to Chapter-9. "**Empirical pore pressure methods (Pp = Pn + ΔP) Pore Pressure Prediction Techniques**", for further explanation.

Empirical Fracture Pressure Methods - There is a multitude of empirical stress ratio (σ_h / σ_v) vs. depth relationships which were summarized by Pilkington (1978). They are shown on figure-4.1a. figure-4.1b shows an equally scaled compactional strain explanation for the empirical fracture pressure methods.

Rocha and Bourgoyne (1996) have shown some excellent results in determining fracture pressure from overburden (S_v) in the Gulf Coast. An explanation for their accuracy and comparison to the combined First & Second Fundamental *in situ* stress/strain method accuracy is presented in, "Discussion: A New Simple Method to Estimate Fracture Gradient", by Phil Holbrook (1996)

Equivalent Depth Methods- Empirical pore pressure methods that assume the pore pressure in the "overpressured" interval has the same effective stress as the same raw petrophysical measurement on the "Normal compaction" **Pn** empirical relationship. The equivalent depth methods assume that the Overburden Gradient is equal at both depths.

Observed Overburden Gradients almost invariably increase with increasing depth that would make the underlying assumption false. Though based on false assumptions, there are many Equivalent Depth Methods in use today. Refer to Chapter-9. "**Empirical (Pp = Pn + ΔP) Pore Pressure** Prediction Techniques", for further explanation.

Evaporite mineral - a mineral which precipitates during the evaporation of fresh or marine waters. The order of precipitation of evaporite minerals from seawater is the reverse order of their solubility and hardness. Precipitation order is also the order of increasing plastic compressibility of evaporite mineral. Figure-1.2 shows Halite's relationship to the compaction of other sedimentary minerals.

Excess pore fluid pressure - The amount of pore fluid pressure which is greater than that exerted by the column of water extending to the earth's water table or sea level. The empirical pore pressure methods use this as (Δp) in their P = Pn + ΔP approach.

Extended Elastic Equations – A group of elastic wave equations that relate V_p^2 and V_s^2 to mineralogy from 0 to 100% porosity. The Gassmann (1951) equations, Woods equation and Hashin-Schtrikman (1963) equations, and Archer's (1992) NaCl brine relationships are all forms of

Hooke's law. Holbrook, Goldberg & Gurevich, (1999). figure-5.2 shows
the **Extended Elastic Equations** mechanical systems domain.

First Fundamental in situ Stress/Strain Relationship - The *in situ*
force balanced power law linear compactional relationship for a granular
solid. The two compaction coefficients in this relationship are
mineralogic material properties, α and sigma max (σ_{max}). For further
information see Chapter-1.0 **THE GOVERNING PHYSICS OF
GEOPRESSURE IN THE SUBSURFACE.**

Fluid - Gas or liquid which is free to flow under pressure.

Fluid Expansion excess pore pressure mechanisms - Any one of
several thermal or thermochemical reactions that results in a net volume
increase of fluid in the pore space of a sedimentary rock. The thermal
fluid expansion coefficient of water is very low.

A highly effective low permeability - high fracture propagation pressure
seal is usually required for the rate of aquathermal fluid expansion to be
greater than the rate of fluid escape. The solid to liquid to gas phase
change of hydrocarbons that results from increasing temperature can
result in significantly large net fluid expansion.

Refer to Chapter-7. FRACTURE PRESSURE LIMIT TO *in situ* PORE
PRESSURE GRADIENTS. Refer to chapter-8. For effective stress un-
loading limb calibration to pore pressure from the fluid expansion
pressurization mechanism.

Force Balance Definition - A physical-mathematical equation defining a
force balance relationship in the Earth. The effective stress theorem,
equation-1.2 is a global boundary condition for granular solids.

The Terzaghi Effective Stress "Law" is a gravitational force balance
definition that applies to biaxial Normal Fault Regime Basins. Horizontal
effective stress is un-defined in the Terzaghi "Effective Stress Law" and
it makes no reference to strain.

The effective stress theorem , equation-1.2 is combined with a definition
of strain. Terzaghi's "Effective Stress Law" applies coincidentally in
biaxial Normal Fault Regime basins. Chapter-6. and figure-6.1 show the

complete closed-form force balanced definition for Normal Fault Regime basins with accompanying explanations.

Fracture Gradient - The borehole fluid pressure divided by the depth of that point necessary to open or propagate an existing fracture in a borehole. The units are in density.

Fracture Initiation Pressure - The borehole fluid pressure necessary to open a new fracture in an unfractured borehole or rock.

Fracture Permeability - The Darcy law resistance to flow through a fracture. Fracture opening width and fracture permeability is highly sensitive to the effective stress exerted against the fracture walls by the fluid pressure within the fracture. Figure-7.1, Figure-7.2, and Figure-7.3 and accompanying force balance self-regulating text provide an in depth explanation.

Fracture Propagation Pressure - The borehole fluid pressure necessary to open an existing fracture in a fractured borehole. Chapter-7. describes the regulating control of fracture pressure over pore pressure profiles.

Free Water Level - The lowest level in a hydrocarbon-containing reservoir that contains any separate fluid phase hydrocarbons. Below this level only single phase free water exists. This is also the minimum work leakpoint of a pressure compartment caprock.

Grain - A small solid particle with a recognizable outline. Abrasion occurring during transport usually rounds natural sedimentary grains. Fossil hard parts also become particles have recognizable shape generated by the once living organism. Most sedimentary grains are composed of a single mineral.

Grain Matrix Framework - The inter-connected solid portion of a porous granular solid - fluid mixture that bears the **effective stress** load and experiences volumetric strain in response to changes in effective stress.

Granular Solid - A body of solid matter composed of grains that are in physical contact and can support a shear wave.

Halite - NaCl , hardness = 2.5 , solubility = 350000 ppm @STP. The most common water soluble mineral. The third evaporite mineral to precipitate from the evaporation of average seawater. Halite often forms extensive salt beds. Salt beds deform easily to form salt domes.

Halite is the highest plastic compressibility evaporite mineral. The precipitation order of evaporite minerals from normal salinity seawater is also their order of decreasing hardness, increasing solubility and decreasing compaction resistance (σ_{max}). See table-1.1 , figure-1.2 , and chapter-1.0. for comparisons.

Horizontal - A direction perpendicular to vertical. Horizontal is parallel to the surface of a body of liquid on Earth.

Hydraulic Potential - The difference between actual fluid pressure and the pressure of a hydrostatic column at the same vertical depth. Fluid will always flow from higher to lower hydraulic potential. Darcy's law gives the relationship by which non-turbulent fluid flow velocity can be calculated.

Hydraulic Potentiometric Field- The three dimensional representation of the difference between actual fluid pressure and the pressure of a hydrostatic column at the same vertical depth. Fluid will always flow from higher to lower hydraulic potential. The direction of fluid flow will be perpendicular to planes of equal hydraulic potential. Darcy's law gives the relationship by which non-turbulent fluid flow velocity can be calculated.

Hydrostatic Pressure - The pressure exerted by a continuous column of fluid according to its average ($\rho \cdot g \cdot h$) below sea level. Hydrostatic pressure is the **Pn** reference pressure in the Empirical **P=Pn+ΔP** Pore Pressure Predication Methods. Figure-9.1 in chapter-9. show the many related empirical observations.

In situ - In place and under natural conditions of temperature, pressure, stress and loading rate. A special natural case of material and measurement of physical properties.

In situ Rock Mechanics System - the static equilibrium force balanced relationships that are determined from in situ petrophysical and borehole fluid pressure measurements. Comparison and discussion of the *in situ* Rock Mechanics System determined relationships to Laboratory Rock Mechanics Systems is given in the article "The use of petrophysical data for well planning, drilling safety and efficiency", by Holbrook (1995) relates the measurements from the borehole and laboratory mechanical systems.

Initial porosity - The porosity of a freshly deposited sediment on the seafloor which has not experienced any additional confining stress . Shumway (1960) figure-2.5. , showed that the porosity of sediment in the first 10 cm of the seafloor is a strong function of average grain particle size in natural quartz - clay mixtures. This relationship closely approximates critical porosity and is the Wood's equation limit that is shown on figure-5.2.

Initial Shut in Pressure - The first pressure measurement made in a cased borehole. ISIP's are routinely made at the start of production to estimate and forecast production rates. It is a direct measurement of the in situ pore pressure.

Laboratory Rock Mechanics Systems - Any rock sample physical properties measurement system which uses a load cell with accompanying **strain** and **fluid pressure** gauges to control or monitor the behavior of a granular solid or rock in a laboratory. The article, "The use of petrophysical data for well planning, drilling safety and efficiency", by Holbrook (1995) compares the comparable force balanced data input control for the laboratory and In situ Rock Mechanics Systems.

Leakoff test - A borehole pressure measurement which determines the break from linearity of a series of pressure vs. volume pumped measurements. The test is normally done after setting casing. The lower pressure linear relationship represents the linear compression of the borehole fluid and elastic expansion of casing and borehole. The leakoff pressure is the decreasing point of departure from the linear trend indicating that additional fluid is being lost downhole. See Kuntz & Steiger (1992) for explanation.

Limestone - A sedimentary lithotype composed almost entirely of calcite with possibly minor amounts of dolomite, clay and quartz.

Liquid limit - The water content, in percent, of a soil at an arbitrarily determined boundary between the liquid and plastic states. ASTM test D 4318 - 84. Note 2- The undrained shear strength of a soil at the liquid limit is considered to be 2 + or - 0.2 KPa (0.28 psi). The uppermost foot of freshly deposited marine sediment is approximately at the liquid limit.

Lithotype - A rock name which characterizes the mineralogic composition.

Loading Limb - The **stress/strain** path relationship of a material that is under increasing load conditions. Whether loading is cyclic or not, any increasing load which is above the previous peak load is on the loading limb. Explanation of loading limb can be found on figure-8.1 in chapter-8.

Lost Circulation - The lack of return of borehole fluid in the annulus that is pumped down the inside of a drillstring. This is almost always an indication of massive fracturing in the open borehole. The volume of fluid pumped and not returned is almost certainly opening and entering the fracture system. Lost circulation occurs when the borehole fluid pressure grossly exceeds fracture propagation pressure.

Marl - A sedimentary lithotype composed of about equal amounts of calcite and clay.

Maximum principal stress - The greatest of the three earth's principal stresses; vertical or horizontal.

Mineral - Any naturally occurring homogenous inorganic substance having a definite composition and characteristic crystalline structure, color and hardness. Minerals are the solid component of the earth mechanical system.

Mineralogic mixing law - The loading limb effective stress/strain (solidity) relationship for a mixed mineral sedimentary rock is the volume weighted average of the single mineral compaction coefficients (σ_{max}) and (α). Equation-1.3 is the First Fundamental grain-matrix-compactional mixing law that is explained as steps on page 28.

The density and elastic coefficient mixing laws for end-member **minerals** and fluid are also linear on the V_p^2 vs. V_s^2 plane. For additional information see; the Ternary mineralogic diagram Figure-2.7, Pressure solution diagram Figure-2.2 a. , binary mineralogic mixing law diagrams Figure-3.4 , Figure-3.5 , and Figure-3.6 , and Chapter-3.6 .

Minimum principal stress - The least of the three earth principal stresses; vertical or horizontal. See figure-4.3 for visualization of Andersonian classification of principal stresses and fault regimes.

Minimum Work Leakpoint - The location along the boundary of a continuous pressure compartment that can be most easily breached by the fluid pressure within that compartment. Capillary pressure must additionally be overcome if a two phase fluid is present.

Figure-7.2 shows the geometric relationship of the Minimum Work Leakpoint with respect to a pressure compartment. A thorough explanation is given in Chapter-7. , "Force Balanced Regulation of Compartment in situ Pore Fluid Pressure by Sealing Caprock in situ fracture pressure".

Normal Fault Regime - An Earth stress state where the maximum principal stress is vertical. The maximum external confining stress is overburden. See figure-4.3 for visualization of the Andersonian relative stress magnitude classification and relative fault displacements.

Open borehole - The lowermost portion of a borehole in the earth which is not separated from the earth by protective casing or liner. The fluid pressure in the open borehole is in direct contact with the earth's in situ pore fluid pressure and fracture pressure.

Overburden - The weight of solid and fluid above a particular point in the subsurface. Also, the vertical component of confining stress.

Overburden Gradient - The weight of solid and fluid above a particular point in the subsurface divided by the depth of that point. The units are in density.

Overlay Methods- Empirical pore pressure methods which estimate the pore pressure from a projected "Normal compaction" **Pn** empirical relationship. The scaling of the empirical ΔP relationship is made directly to measured pore pressure in the "overpressured" zone. The empirical ΔP relationship is only a function of depth below the base of the "Normal compaction" **Pn** empirical relationship. The scaling of the raw measured parameter empirical ΔP relationship is usually different between the effective stress scaling in the "Normal compaction" **Pn** empirical relationship.

Pascal's principle - If added pressure is applied anywhere to a confined fluid, that pressure is transmitted undiminished to every portion of the fluid and to the walls of the confining vessel. The pressure difference between two points in a connected fluid is determined only by the difference in <u>vertical</u> height the fluid <u>density</u> and the earth's gravitational field. Blaise Pascal (1623 - 1662).

<u>Permeability</u> - The property that permits the passage of fluid through the interconnected intergranular pores or fractures in a rock. The intergranular permeability has a very low sensitivity to <u>effective stress.</u> <u>Fracture permeability</u> is highly <u>stress</u> sensitive. See <u>chapter-7.</u> that explains the dynamic relationship of net (fracture + intergranular) permeability.

<u>Plastic</u> - An irreversible **stress/strain** relationship wherein the sample being tested does not change it's final dimension when the <u>stress</u> is removed. A material capable of undergoing continuous deformation without rupture or relaxation.

Owing to pressure solution mass transport that occurs over geologic time; <u>elastic</u> minerals in a <u>grain matrix framework</u> exhibit dominantly plastic deformation. See <u>Figure-2.2</u> a for a graphic representation of irreversible pressure solution consolidation. First Fundamental in situ Stress/Strain Relationship is a <u>plastic</u> power-law linear mathematical function (see figure-<u>1.2</u> , <u>chapter-1.0</u> and <u>chapter-2.0</u>)

<u>Pore pressure</u> - The <u>fluid pressure</u> within the pore-space of a <u>granular</u> <u>solid</u>.

<u>Porosity</u> - The fractional volume of a body of matter which is not <u>solid</u>. The mathematical complement of porosity (ϕ) is <u>solidity</u> (1.0 - ϕ).

<u>Porous granular solid</u> - A porous material initially composed of separate <u>solid</u> <u>grains</u>. The <u>grains</u> must be in contact forming a framework which can resist shear <u>stress</u>. The <u>grain</u> contacts can be <u>cement</u>ed or sutured during consolidation reducing <u>porosity</u>. Some measurable <u>porosity</u> is required for this definition.

<u>Power-Law Relationship</u> - a linear relationship between the logarithms of two different variables. <u>Equation-1.3</u> represents the power-law linear First Fundamental in situ Stress/Strain Relationship. <u>Figure-1.2</u> , shows that relationship for the common sedimentary minerals. <u>Chapter-1.0.</u> explains the relationships to the governing physics. The Extended Elastic Equations of Hooke's law are a continuous $V_p^2 - V_s^2$ power-law relationship, see <u>figure-5.2.</u>

Pressure Integrity Test - A series of borehole pressure vs. volume pumped measurements which do not break from linearity. The test is normally done after setting casing to determine if the open borehole just below the casing shoe is strong enough to contain the maximum borehole fluid pressures expected in the next planned open borehole interval. The PIT provides a minimum value of fracture pressure at the casing shoe.

Pressure Compartment - A moderate to high permeability continuous lithostratigraphic body that has equilibrated to seal relative hydrostatic pressure over geologic time. The general seal - pressure compartment earth force balanced Darcy law fracture pressure/pore pressure inter-relationship is summarized below.

Chapter-7. gives a thorough explanation of the interdependent force balanced regulation of compartment pore pressure. The fracture propagation pressure of the pressure seal at the minimum work leakpoint determines the maximum seal relative hydrostatic pore pressure within a pressure compartment.

Pressure Seal - A low intergranular permeability lithotype which also has a sufficiently high horizontal stress to hold subvertical fractures closed. The pore pressure seal will be broken if the pore pressure within or immediately below the seal reaches the fracture propagation pressure of the seal. The general seal - pressure compartment earth force balanced Darcy law fracture pressure/pore pressure inter-relationship is explained in Chapter -7.

Pressure solution - An atomic mass transport process that preferentially dissolves ions within an existing mineral lattice framework at points of locally higher stress. The dissolved ions often migrate along recognizable intergranular films and then precipitate in an intergranular void that is under locally lower stress. Pressure solution tends to redistribute load bearing matter and stress more uniformly within a sedimentary rock. Figure-2.2 a shows a graphic representation of consolidation and accompanying pressure solution.

 The bond strength and load bearing capacity of a mineral is the same whether it occurs as a recognizable grain or cement.

Quartz - SiO_2, Hardness = 7, Solubility @STP= 6ppm. The most common mineral composing 52% of the Earth's crust. Also the hardest and most weathering resistant of the common minerals.

Repeat Formation Test - A wireline conveyed pressure transducer that isolates itself from the borehole fluid pressure by pressing a hollow probe against the formation wall. The device can take many fluid samples from separate formations. Measurements from this device are often called RFT's. It is a direct measurement of the in situ pore pressure.

Safe Drilling Window - The range of borehole fluid pressure gradients between the maximum pore pressure within the open borehole and the minimum fracture propagation pressure. Refer to Safe Drilling Window in "The Use of Petrophysical Data for Well Planning, Drilling Safety and Efficiency" by Phil Holbrook (1995), for further explanation.

Sandstone - A sedimentary lithotype composed dominantly of quartz sand size grains. This lithotype may contain calcite or clay as minor constituents.

Seal Relative Hydrostatic Pressure - The pressure exerted by a fluid within a sealed pressure compartment that at any point is equal in all directions. The pressure difference within a continuous pressure compartment can be calculated from fluid density using Pascal's principle. Equation 7.1 gives this relationship with respect to the minimum work leakpoint. The constant pressure offset for the compartment is the minimum fracture propagation pressure of the seal. Figure-7.3 and Figure-7.5 show the fracture pressure seal relative hydrostatic pressure relationship for many pressure compartment s.

Second Fundamental in situ Stress/Strain Relationship - The in situ force balanced linear minimum/maximum effective stress ratio to strain (solidity) relationship for porous granular solid in Normal Fault Regime biaxial basins. The two compaction coefficients in this relationship are mineralogic material properties, α and sigma max . For further information see, figure-4.1a , Figure-4.1b , and Figure-4.2 in Chapter-4.0

Self regulating - The property of a system that tends to resist change and keep it within a certain range of states. In the subsurface caprock fracture pressure is a **force balanced self regulating** upper limit to underlying continuous pressure compartment pore pressure. Chapter 7. gives a thorough explanation of the interdependent force balanced regulation of pressure compartment pore pressure.

Sedimentary Rocks - A solid assemblage of mineral grains that was deposited by the gravitational settling onto the surface of the Earth. If the earth surface is above sea level these would be deposited from air and be termed "aeolian" deposits. Sediments settling through water to form lake deposits are called "lacustrine". Sediments deposited on the sea floor are called "marine".

Sigma max - The total compaction intercept (solidity = 1.0) σ_{max} of the in situ force balanced power-law linear relationship for a granular solid. Sigma max is one of two mineralogic compaction coefficients in First Fundamental in situ Stress/Strain Relationship. (see chapter-1.0 and figure-1.2). It is also the power law intercept of the Second Fundamental in situ Stress/Strain Relationship. (See figure-4.2).

Shale - Very fine grained sedimentary deposit composed dominantly of clay minerals. Quartz and calcite are usually the minor constituent in shales composing 10% to 40% of the solid volume. Shales compose about 70% of all lithotypes in sedimentary basins. However, the volume of clay minerals in a shale lithotype is often less than 30%.

Solid - Matter that contains an internal bond structure that can resist shear stress.

Solidity - The fractional volume of a body of matter which is solid. The mathematical complement of solidity ($1.0 - \phi$) is porosity (ϕ). Figure-1.1 shows the relationship of solidity to absolute in situ Strain.

Static equilibrium - A steady state where both solids and fluids are essentially at rest each bearing its fractional load proportionally. The effective stress theorem , Equation-1.2 is the boundary condition describing the load distribution at the static equilibrium state.

Strain - The relative change in dimension (length or volume) when an external confining load is applied to a body of matter. Solidity is by definition the inverse of volumetric strain for porous granular solids that has a fixed upper limit of 1.0. Figure-1.1 shows the grain-matrix-compactional strain relationship to solidity and porosity.

Stress - A load applied in a direction to a solid body.

Strike-Slip Fault Regime - An Earth stress state where the intermediate principal stress is vertical. See figure-4.3 for visualization of the Andersonian relative stress magnitude classification and relative fault displacements.Equations-4.3 substituted into Figure-6.1 express the closed form force balance in this tectonic regime.

Terzaghi Effective Stress Law - Uniaxial gravitational force balance definition stating that confining stress on a body of matter; is borne Overburden, the vertical by the sum of the pressure plus the vertical effective stress (Terzaghi, 1923). The article, "A simple closed form solution for Pore Pressure, Overburden and the principal Effective Stresses in the Earth", by Phil Holbrook (1999) relates the uniaxial Terzaghi Effective Stress Law to the global volumetric Effective Stress Theorem.

Thrust Fault Regime - An Earth stress state where the minimum principal stress is vertical. This is the third Andersonian relative-stress-magnitude classification related to relative fault displacements.

True vertical depth - The vertical difference in elevation between a datum near the surface and a point in the subsurface. The most common datum's are the drill floor, Kelly bushing, sea level, surface elevation or mud line. Sea level is the most significant datum for offshore locations as overburden starts at sea level.

Two phase fluid - A fluid composed of two immiscible fluids. For example, water and oil, or water and gas.

Un-loading Limb - The **stress/strain** path relationship of a material that is under decreasing load conditions. Whether loading is cyclic or not, any decreasing load which is below the previous peak load is on the un-loading limb. Depending on geologically slow associated chemical reactions; an un-loading limb can range from purely elastic to partially plastic. See chapter-8. and figure-8.1 for the stress/strain path α_{offset} relationships.

Vertical - The direction away from the center of the Earth.

Water Conductivity Profile - The change in formation water conductivity that usually occurs uniformly with depth in a well or local area.

About the Author

Phil Holbrook received his Ph.D. in Geology from Penn State in 1973. He entered Gulf Oil Company's one year Experience Broadening program in Exploration Geophysics. Gulf's EB program is equivalent to an M.S. in Exploration Geophysics. He then worked for 2 years in exploration operations, interpreted seismic data, developed drilling prospects, and supervised acquisition of geophysical data and well logging.

He transferred to Gulf's R&D where he then coordinated Gulf's EB program. There he also conducted research on petrophysical oil exploration applications. He worked at Exxon Production Research for 8 years performing reservoir studies, seismic well ties, and ran borehole televiewer operations. He also served as an internal well log analyst-petrophysicist consultant to several of Exxon's operating divisions.

He joined Sperry-Sun in 1984 and began work on a mechanically representative pore pressure and fracture pressure prediction methodology in 1985. This physically representative methodology has been developed and applied to over 300 wells worldwide. Pore Pressure through Earth Mechanical Systems explains the relationships from 15 years experience and mechanical intuition. The earth mechanical systems relate porosity and mineralogy <to> Newton's law <to> Hooke's law for in situ sedimentary rocks.

He has studied geology, petrophysics and earth mechanics and written over 25 technical papers in these areas. Presently he consults on petrophysics, earth *in situ*, and borehole relative force balance. He performs applications research projects; and teaches subjects that are important to subsurface engineering and earth science.

Pore Pressure through Earth Mechanical Systems will be updated periodically depending on new information and reader feedback. Immediately below are the scientific and practical reasons for understanding earth mechanical systems. You can go to chapter-1.0 to start at the beginning. Please feel free to comment by e-mail, fax, or telephone call. Phil Holbrook can be contacted pholbrok@flash.net or calling 713-977-7668, fax 713-977-0477.

Serious Reading for Geologists and Earth Subsurface Engineers

A mechanical systems hypothesis for the earth's interior has not been previously advanced. The hypothesis put forward is that the earth is a classic mechanical system wherein forces are balanced and effective stress is everywhere proportional to *in situ* strain. Stress and strain are mathematically proportional in the earth mechanical system domains.

Pore pressure is part of the earth's coupled fluid-solid mechanical system. Each constitutive mechanical system domain is described in terms of fundamental rock properties (ie.) porosity, mineralogy, and pore fluid composition. Chapter 6 relates the limiting extended-elastic-equations and grain-matrix-compactional mechanical systems.

Effective stress-stress and effective stress/strain data support constitutive mechanical system hypotheses in both Normal Fault Regime and Strike-Slip fault regime basins. These relationships are shown and explained in chapter 4 and chapter 7.

There is measurable hysteresis in the mechanical behavior of *in situ* granular solids as there is in laboratory load cells. Chapter 8 describes the nature and calibration of geologic time scale unloading-limb mechanical systems.

Hooke's law relates dynamic and static elastic mechanical systems. Hooke's law has been extended from flawless mineral solids to clear NaCl brine fluids in a continuous closed-form set of equations. The static form of Hooke's law is a limit in both laboratory and natural subsurface stress/strain relationships. The dynamic form of Hooke's law governs both borehole and seismic elastic wave propagation. Hooke's law is a cross discipline physically representative tie-point between borehole mechanical and remotely sensed acoustic rock properties.

Newtonian gravitation, Hooke's law, and Pascal's principle have been combined with a definition of absolute *in situ* strain. This synthesis describes the force balanced physics of the earth's sedimentary crust. The explanations and proofs of these mechanical systems relationships are organized as chapters in this book.

It is sincerely hoped that earth mechanical systems will be as informative and practical when applied within the earth as Newton's, Hooke's, and Pascal's mechanical systems. Earth mechanical systems are a constitutive, mathematically simple extension of the previous governing physics.

About the book cover

The upper block diagram on the cover represents fluid-filled borehole in the earth. The open borehole drilling limits are Pore Pressure (P_p) and Fracture Propagation Pressure (P_f). Borehole fluid pressure (**Pb**) opposes pore pressure within the earth's fracture pressure limits.

Pore Pressure and Fracture Propagation Pressure are naturally balanced against each other through the earth's coupled mineral-fluid mechanical system. The opposing fracture and pore pressure symbols →← on the block diagram portrays this force balance. Borehole fluid pressure (**Pb**) must be kept within the earth's pore pressure and fracture pressure limits for safe, trouble free drilling.

The chapters in this book explain how the natural drilling limits are related to each other through the earth's governing physics. Newton, Hooke and Pascal separately described the governing physics in the mid-1600's. They are honored with their pictures on the book's cover. Their scientific discoveries are synthesized with strain using the most fundamental earth material properties (porosity, mineralogy, and pore fluid composition).

Hooke never allowed a portrait of himself to be painted possibly due to a spinal deformity. An image of Yoda was substituted because both Hooke and Yoda were able to perceive and forecast events across time and space. Hooke's law relates elastic stress to strain and is with us in every man made and natural object we see. Through extension to clays and fluid, Hooke's law now is fully implemented in the earth's sedimentary crust.

Over 98% of the raw information that we have about the earth today was derived from borehole petrophysical sensors and reflected seismic waves. This raw information is synthesized under both Hooke's law and Newtonian gravitational domains. Mountains of raw data are thereby explained through a few simple physical equations that represent the fundamental governing physics. This is a revolutionary physical alternative to the 250+ empirical methods that have been directed toward the earth's pore pressure P_p and fracture pressure P_f in the last 35 years.

The entreatment, "Let the Petrophysics balanced Force be with you", guides earth scientists toward nature's mechanical path to interpret today's remotely sensed data. Forces are balanced in the earth and stress is everywhere proportional to strain. These mechanical system tenets have been used successfully for centuries. This book shows that the same governing physics applies below the earth's surface.

The ⎡inset box⎤ at the bottom of the cover summarizes the scientific synthesis. Force balanced boundary conditions apply in conjunction with a definition of absolute *in situ* strain. The binding of Newton's law, Hooke's law and Pascal's principle to earth physical properties offers a mathematically simple solution to many practical earth science problems. The Force↔Balanced Physics of the Earth's Sedimentary Crust is explained in this book.

www.ingramcontent.com/pod-product-compliance
Lightning Source LLC
Chambersburg PA
CBHW070732220326
41598CB00024BA/3397